パーフェクト獣医学英語

Perfect Veterinary English Terminology

谷口和美 著

チクサン出版社

はじめに Preface

　本書は、獣医学用語を語源に触れながら接頭辞、接尾辞、語幹に分け、解剖学から臨床にわたる幅広い知識を分かりやすく組み立てたものである。動物にかかわる英語を勉強している人、動物にかかわる仕事をしている人、とりわけ獣医学科の学生、獣医師、動物看護学や畜産学などを学ぶ方のための本である。

　専門知識や新しい知見の多くが英語で発表されている現在、その知識の不足は、情報収集能力の差につながり、大きなハンディキャップになってしまう。とりわけ獣医学・医学用語は数が多いので、系統的に学ぶ必要がある。いわゆる西洋医学はギリシャ、ローマ時代にその源流をたどることができる。そこで用いられた医学英語もまた、修得のためにはギリシャ文明やラテン文明にさかのぼることが肝要となる。

　日本に西洋医学がどっと入ってきたのは、江戸時代末期、明治維新直前である。解体新書のころ、杉田玄白らは、西洋の解剖学用語を何とかして日本の言葉に置き換えようと、大変な苦労をされた。そのとき、手持ちの語彙の中で、邦訳した結果、同じ漢字を何度も繰り返し使うという現象がおこった。

　例えば、胸骨（sternum）を漢字でとらえると、胸郭（thorax）と骨（bone）がイメージされるが、英語は sternum であって thorax-bone ではない。このように、日本語の解剖学用語だけを断片的に理解しても、英語の専門書は読めない、英語での議論もできない、ということにもなりかねない。どうしても、英語は系統的に学ぶ必要がある。

　読者諸氏には是非、断片の集積でない、獣医学英語の世界のイメージを体系的に理解していただきたい。そう、心から祈っている。

定冠詞の取り扱い

　本書の臓器名などの多くで、定冠詞（the）を省いてあるが、原則は臓器名には定冠詞を付ける。例えば心臓は heart ではなく the heart であり、肺は lung ではなく the lungs である。前者が単数形なのは心臓が動物当たりひとつしかないからで、肺は右肺と左肺の2つあるため複数形となる。She has heart. という定冠詞を省いた文は、「心臓がある」ではなく「勇気がある」という意味である。

　本書の校正段階で、定冠詞、単数形、複数形の扱いには大いに悩んだ。結果、分かりやすさを第一に、煩雑な定冠詞を省いた。現場では場面に応じて、臓器名には定冠詞を付け、単数か複数かを考えながら、使っていただきたい。

2009年　谷口 和美

目次

はじめに ··· 2

Chapter 1　獣医学英語の基本構造 Structure of Veterinary Medical Terms ············· 5
1. 獣医学英語の成り立ち ·· 6
2. 体の向き、位置を表す用語 ·· 8
3. 獣医学用語と医学用語の違い ·· 9
4. 体の各部位 ··· 10
5. 足の方向 ·· 11
6. 骨格系と内臓器官 ·· 12
7. 英語とラテン語の関係 ··· 14
8. 人と動物の英語名 ·· 16

Chapter 2　接頭辞 Prefixes ·· 17
1. 接頭辞 ··· 18
   Exercises ① ·· 27

Chapter 3　接尾辞 Suffixes ·· 29
1. 接尾辞 ··· 30
   Exercises ② ·· 39
2. 炎症 ·· 42
3. 痛み ·· 45
   Exercises ③ ·· 47

Chapter 4　細胞の基本構造 The Structure of the Cell ··· 49
1. 細胞と組織 ··· 50
2. 遺伝学 ··· 53

Chapter 5　筋骨格系とその疾患 The Musculoskeletal System and its Disorders ·········· 55
1. 骨格系 ··· 56
2. 関節 ·· 61
3. 筋 ··· 64
   Exercises ④ ·· 67

Chapter 6　循環器系とその疾患 The Circulatory System and its Disorders ················ 69
1. 循環器系 ·· 70
2. リンパ器官 ··· 82
3. 免疫学 ··· 85
   Exercises ⑤ ·· 87

🔊 マークについて・・・付属CD-ROMに音声データが収録されています。

Chapter 7 呼吸器系とその疾患 The Respiratory System and its Disorders ……… 89
    1　呼吸器系 …………………………………………………………… 90
       Exercises ⑥ …………………………………………………… 96

Chapter 8 消化器系とその疾患 The Digestive System and its Disorders ……………… 97
    1　消化器系 …………………………………………………………… 98
    2　肝臓と胆囊 ……………………………………………………… 115
    3　膵臓 ……………………………………………………………… 119
       Exercises ⑦ ………………………………………………… 120

Chapter 9 尿生殖器系とその疾患 The Urogenital System and its Disorders …………… 121
    1　泌尿器 …………………………………………………………… 122
       Exercises ⑧ ………………………………………………… 126
    2　雌性生殖器 ……………………………………………………… 127
    3　雄性生殖器 ……………………………………………………… 132
    4　臨床繁殖学 ……………………………………………………… 136
       Exercises ⑨ ………………………………………………… 138

Chapter 10 神経系および内分泌系とそれらの疾患 The Nervous and Endocrine Systems and their Disorders … 139
    1　神経系 …………………………………………………………… 140
    2　内分泌系 ………………………………………………………… 144
       Exercises ⑩ ………………………………………………… 146

Chapter 11 感覚器系および外皮系とそれらの疾患 The Sensory and Integumentary Systems and their Disorders … 147
    1　視覚系 …………………………………………………………… 148
    2　平衡聴覚系 ……………………………………………………… 151
    3　嗅覚系 …………………………………………………………… 152
    4　味覚系 …………………………………………………………… 153
    5　外皮系 …………………………………………………………… 154
       Exercises ⑪ ………………………………………………… 156

Chapter 12 腫瘍学、病理学、寄生虫学 Oncology, Pathology and Parasitology ………… 157
    1　腫瘍学 …………………………………………………………… 158
    2　病理学 …………………………………………………………… 160
    3　寄生虫学 ………………………………………………………… 162
       Exercises ⑫ ………………………………………………… 165

    付録
      Ⅰ　専門分野と専門家 ……………………………………………… 166
      Ⅱ　獣医学臨床分野の頻出用語 …………………………………… 168
      Ⅲ　ペットのリハビリ ……………………………………………… 169
      索引 …………………………………………………………………… 171

Chapter 1

獣医学英語の基本構造
Structure of Veterinary Medical Terms

1 獣医学英語の成り立ち Word Structure in Veterinary English

獣医学・医学英語の法則性を知ろう

森鷗外の自伝的中編『ヰタ・セクスアリス』に、若かりし頃の鷗外が医学部で医学用語を学ぶシーンがある。クラスメートたちが単語を丸飲みしようと四苦八苦するのを尻目に、鷗外は単語をいくつかの要素に分解してその成り立ちを考えた。それらの要素の起源をギリシャ語ないしラテン語にたどり、ノートに書き付けるだけで、彼はらくらく理解したのだという。

■ 接頭辞、語幹、接尾辞とは

では鷗外はどのように医学用語を理解したのだろうか？
英語の多くが、**接頭辞＋語幹＋接尾辞**から成っている。もちろんこのすべてを持っているとは限らず**接頭辞＋語幹**、あるいは**語幹＋接尾辞**だけの組み合わせも多いが、この基本構造を理解すると、専門用語に限らず英語の世界全体が広がる。とりわけ獣医学・医学英語は、長く難しい単語が多いが、実は接頭辞、語幹、接尾辞をうまく組み合わせて作られた造語が多い。

接頭辞とは、例えば endocrine（内分泌）、exocrine（外分泌）の endo-、exo- の部分である。これは、endo- が「内側」、exo- が「外側」、という意味であることを知っていると、いっぺんに両方の言葉を理解することができる。それを踏まえた上で次に endoderm（内胚葉）、ectoderm（外胚葉）という発生学用語を目にすると、ああ、あの endo- がまた出てきたな、exo- と ecto- は相同だな、と推測することができるようになる。

接尾辞とは、例えば encephalitis（脳炎）、gastritis（胃炎）、colitis（大腸炎）はすべて -itis で終わるが、この -itis の部分が炎症全般に共通する接尾辞である。このことを知っていると、その次に bronchitis という単語を見て、bronchus が気管支であることが分かっていれば、辞書を引くまでもなく気管支炎だろうと推測できるのである。医学用語には、共通する疾病や術式で同じ接尾辞を持つものが多いため、それらを押さえると効率良く勉強できる。本書では Chapter 2 で接頭辞を、Chapter 3 で接尾辞を学ぶ。

語幹には通常、臓器名や組織名が入る。例えば上記の bronchitis（気管支炎）は bronchus ＋ -itis に分解でき、この場合 bronchus が語幹に当たる。本書では Chapter 5 以降で、体の各臓器、各系ごとに固有の単語を見ていくことにする。
接頭辞、語幹、接尾辞という英単語の構成は、獣医学・医学用語にとどまらず、英語全体に通ずる基本であるので、本書を通読された後、読者諸氏は専門英語だけでなく、一般的な英語の学力も上昇したと感じるはずである。

ギリシャ文明、ラテン文明の影響

　　獣医学・医学英語の起源は、古代ギリシャ、ラテン文明にまでさかのぼる。これに中世、近世のヨーロッパの言葉が加わって、次第に現在のものへと形作られてきた。獣医学や医学は高度に専門的な学問と言えるため、とりわけギリシャ語、ラテン語の影響を今も色濃く残している。例えば英語で肝臓は liver であり、広く一般に知られている言葉だろう。これに加えて、ラテン語で肝臓が Hepar であることを知り得ていると、肝炎がなぜ hepatitis であるかが、突然目の前に広い道が開けるように理解することができる。

　　言葉の成り立ちを知るのは楽しい。もうひとつ例を挙げよう。腎臓は英語で kidney だが、ラテン語は Ren である。そうすると、腎動脈がなぜ renal artery でなければならないかが必然的に理解できる。

　　獣医学・医学用語を学ぶことにより、歴史のなかに脈々とつながれた賢人たちの知識を受け継ぐ感動を味わおうではないか。

日本語の獣医学・医学用語

　　本書の「はじめに」にも書いたとおり、日本語の獣医学・医学用語の多くが明治維新前後の西洋文明の急激な流入時に、無理やり翻訳する形で作られた。手持ちの漢字を組み合わせて造語したため、例えば「舌骨舌筋」は4文字のなかに舌という漢字が2度も出てくる。英語では"hyoglossus muscle"である。翻訳の際、日本語の舌骨（hyoid bone）と舌（glossus）とを、英語の原型は違うもの同士にもかかわらず、それらを無理やりつないだためこうなったのである。

　　専門用語を勉強すると、日本の先人たちが漢文の素養を元に何とか医学用語を邦訳しようと頭を悩ませた様子が見えるような気がする。

　　反対に、何と日本から中国へ"逆輸出"された漢語もある。nerve は「神経」と中国でも訳されるが、これは nerve という概念と共に日本から中国へ渡った言葉だという。日本で創造された漢字さえある。例えば「腺」（gland）は日本が独自に作った漢字だと読んだ記憶がある。

　　ともあれ、舌骨舌筋ではイメージがいまいち希薄であるが、"hyoglossus"と記憶すると、より深くこの筋を理解することができる。専門英語を勉強しながら、一段深い獣医学の知識を身に着けようではないか。

獣医学用語の特異性

　　獣医学用語は、その大部分が医学用語と共通である。しかし、すべて同じというわけではないにもかかわらず、獣医学の専門英語の特異性を述べた類書は著者の知る限り、無い。本書は、一般的な医学用語に加えて、とりわけ獣医師に知っておいてもらいたい獣医学用語を熱を込めて解説した。

　　本書を通読した後には、初めて目にした単語でも既存の知識と組み合わせて、意味を推測することができるようになっているはずである。

　　本書を上手に利用し勉強することで、森鷗外のように悠々と獣医学・医学外国語をモノにしていただきたい。

2 体の向き、位置を表す用語
Anatomical Terms of Orientation

 まずは体の向きを表す用語を理解しよう。

cranial	頭側方の	caudal	尾側方の
rostral	吻側方の		
ventral	腹側方の	dorsal	背側方の
medial	内側方の	lateral	外側方の
intermediate	中間の		
internal	内の	external	外の

　体軸に対して体軸に近いか遠いかを表す用語が medial（内側方の）と lateral（外側方の）である。これに対し、箱のように閉じた構造にとっての内側と外側は、internal（内の）、external（外の）という。

正中矢状断面
Midsagittal Section

median	正中の
sagittal	矢状の
midsagittal	正中矢状の

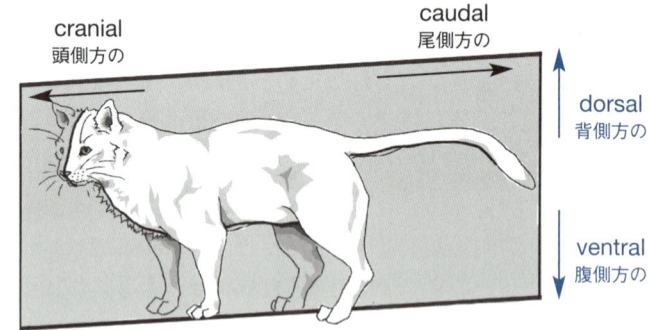

水平断面
Horizontal Section

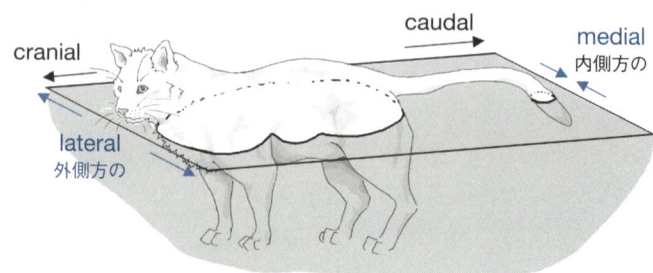

横断面
Cross Section

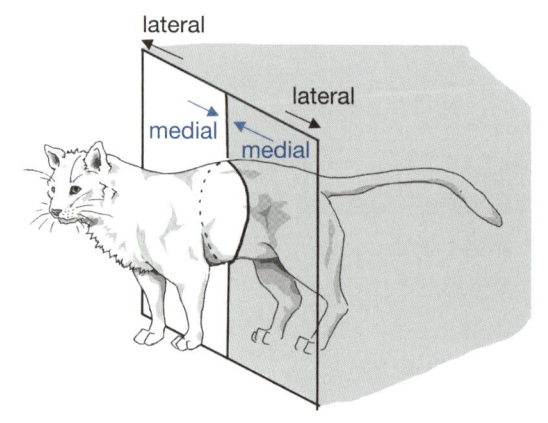

3 獣医学用語と医学用語の違い
Comparison of Anatomical Terms for Humans and Animals

獣医解剖学用語と人体解剖学用語は、方向を表す言葉が異なる。

　獣医学用語の大部分は、医学用語と共通しているが、獣医学用語に固有の言葉もある。その代表例が、体の向きを表す言葉である。

　人間は直立しているため、人体解剖学の用語を獣医解剖学に読み替えようとする時には、必ず体の向きに注意を払い、以下の「読み替え」を行う必要がある。

人体解剖学		獣医解剖学	
anterior	前方の	ventral	腹側方の
posterior	後方の	dorsal	背側方の

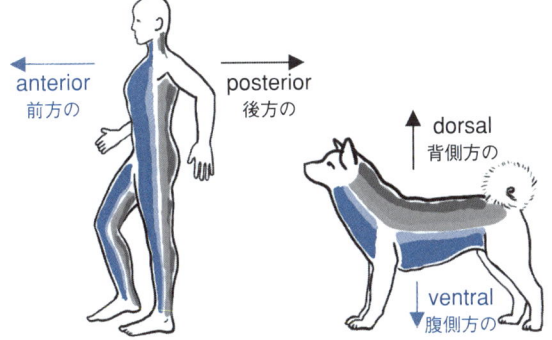

人体解剖学		獣医解剖学	
superior	上方の	cranial	頭側方の
inferior	下方の	caudal	尾側方の

　頭部のなかで、例えば鼻は耳より前にある、という時、鼻も耳も頭部にあるので「頭側方の」という表現は用いない。

　このため獣医学用語では、rostral（吻側方の）という言葉もよく使われる。吻とは鼻先のこと。

　cranial と rostral はよく区別されずに「前方の」という意味でしばしば用いられる。

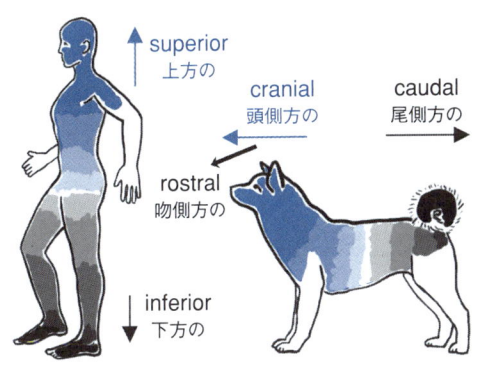

人体解剖学		獣医解剖学	
frontal	前額断の	cross	横断の

　「前額断の」という言い方は、原則として顔面に平行という意味で、人間では顔の面がほぼ垂直に近いため使われるが、獣医学分野では使用頻度が低い。（体幹の横断の時は、人体解剖学も獣医解剖学も cross と言う）

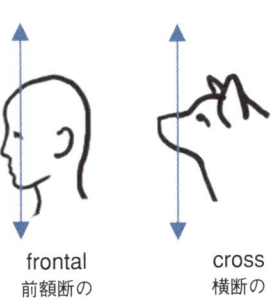

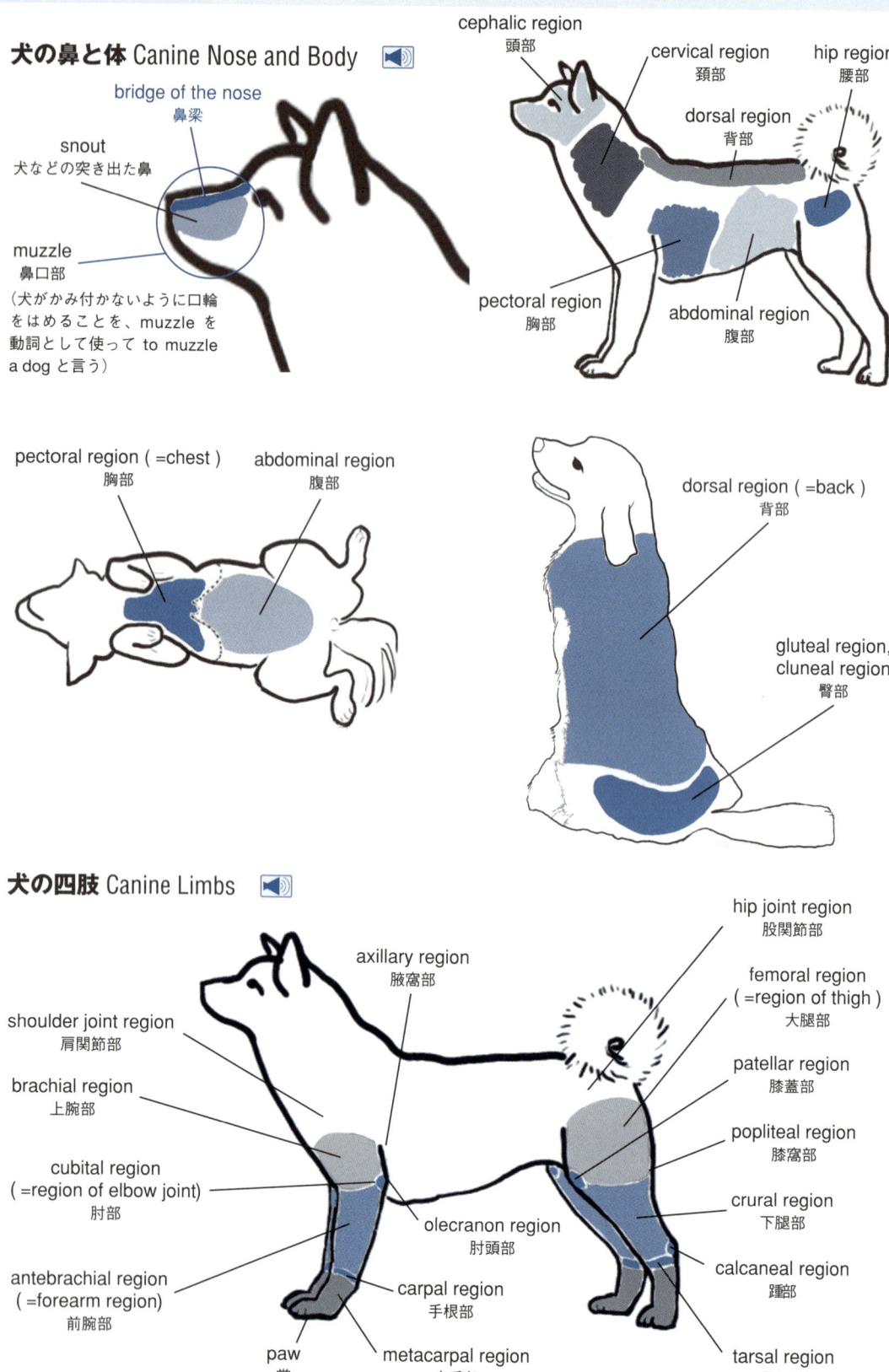

5 足の方向 Positional Terms related to the Foot

方向 Directions

superficial	表層の	profundus	深層の
proximal	近位の	distal	遠位の
peripheral	末梢の	central	中枢の
dorsal	手背側の	palmar	掌側の
dorsal	足背側の	plantar	足底側の

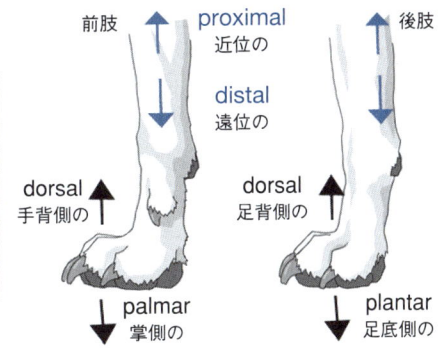

軸 Axis

axis	軸		
axial	軸側の	abaxial	反軸側の

　偶蹄類は第三趾と第四趾を持つ。このように2本以上の指趾を持つ動物で、指間の縦のラインを axis（軸）と言い、軸に近いか遠いかで、軸側、反軸側という。
　これは、正中矢状断面に対して近いか遠いか、という内側、外側という概念と異なる。

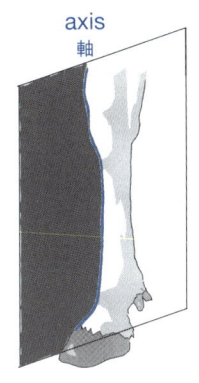

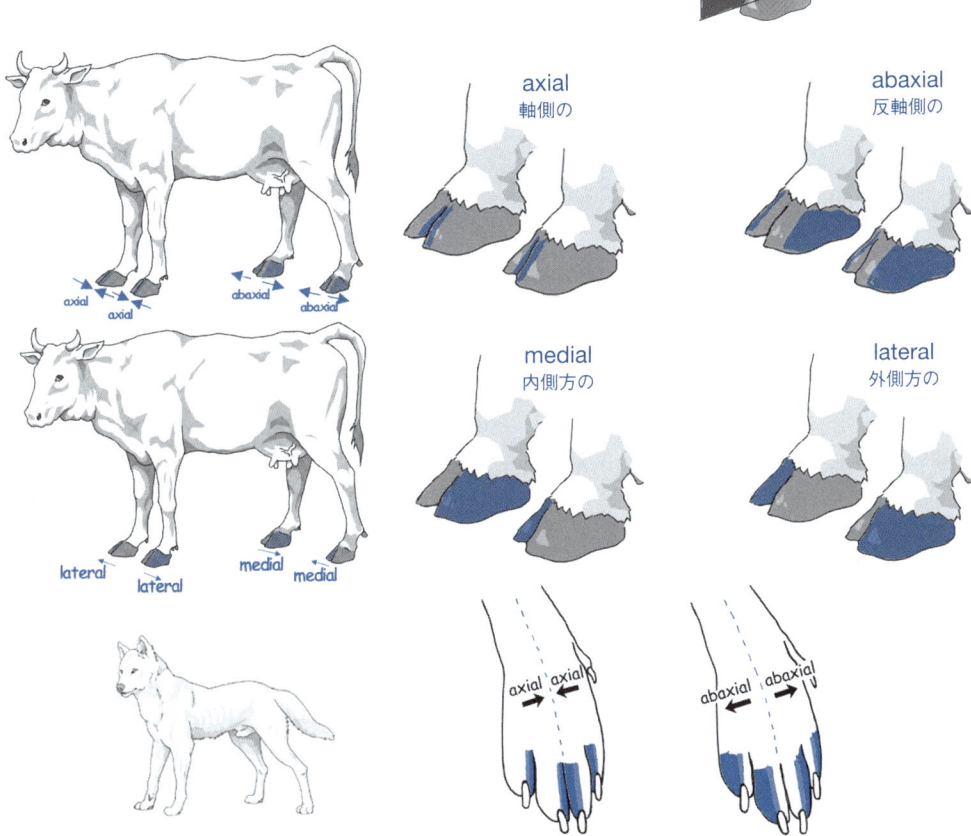

6 骨格系と内臓器官
The Skeletal System and Visceral Organs

体の部位 Body Parts 🔊

骨格系 Skeletal System 🔊

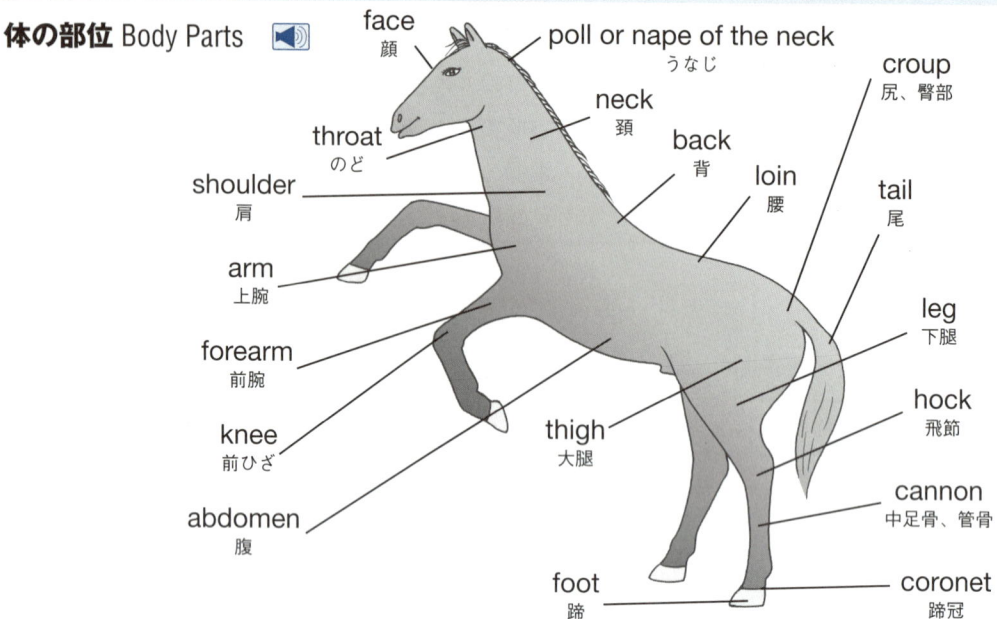

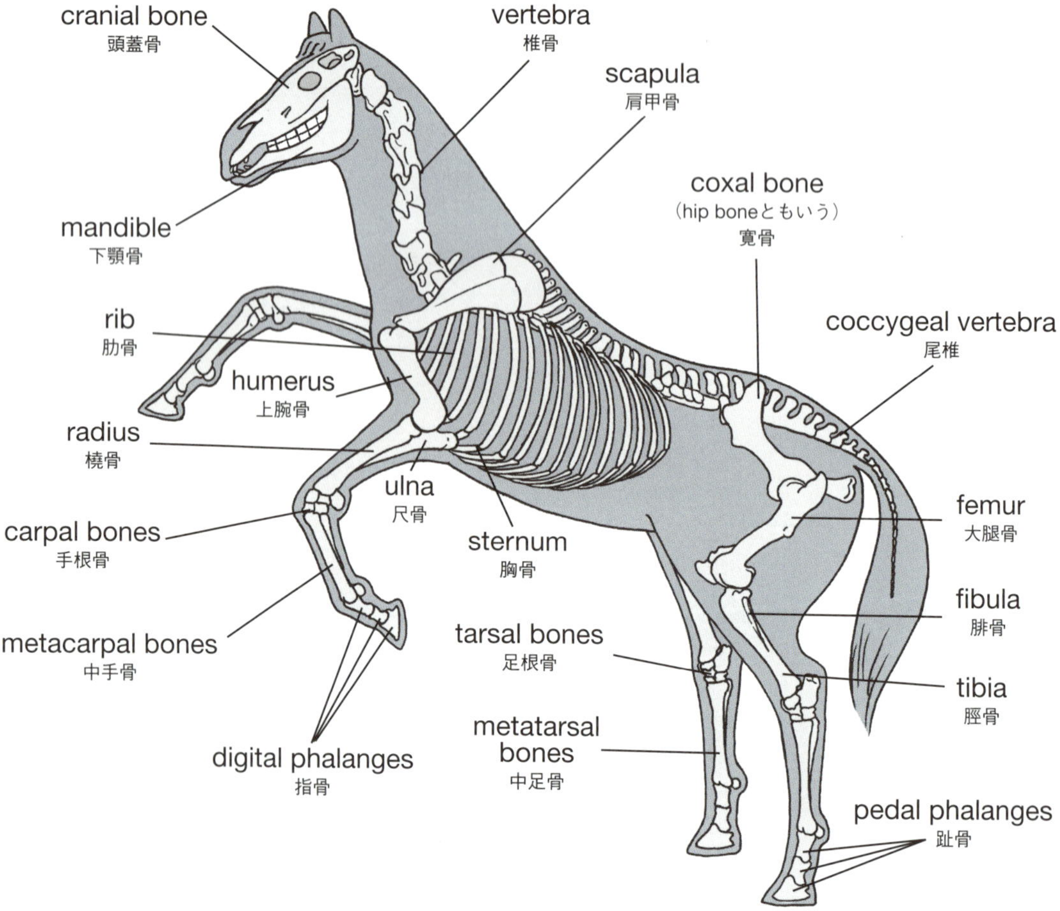

内臓器官（メス）Female Visceral Organs

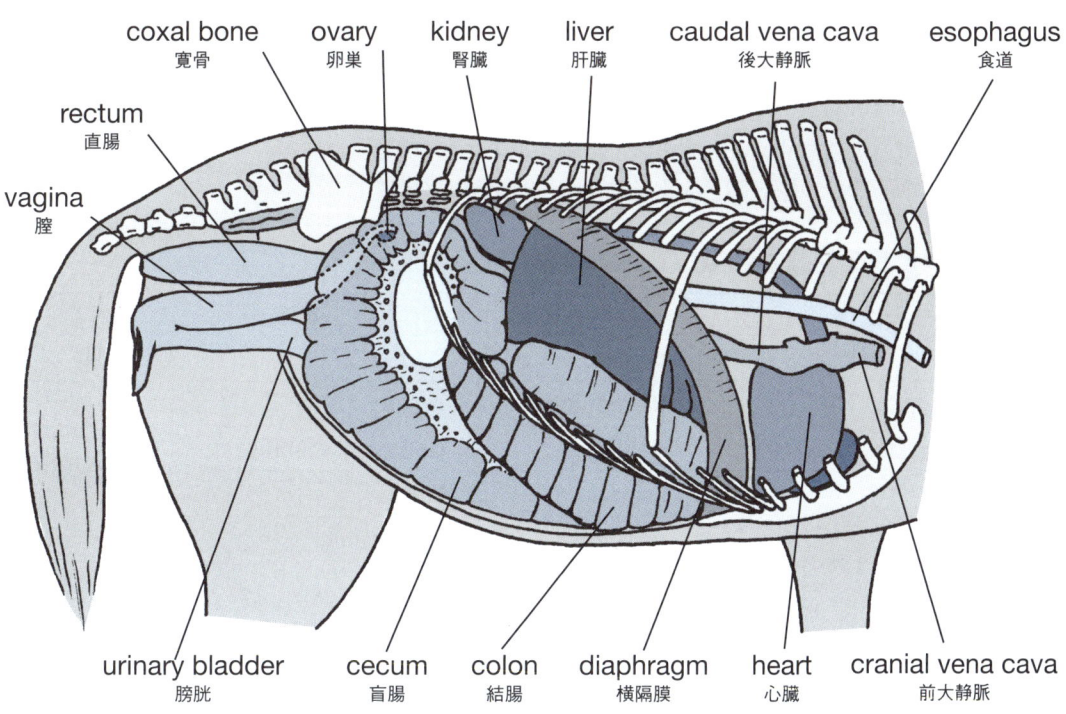

内臓器官（オス）Male Visceral Organs

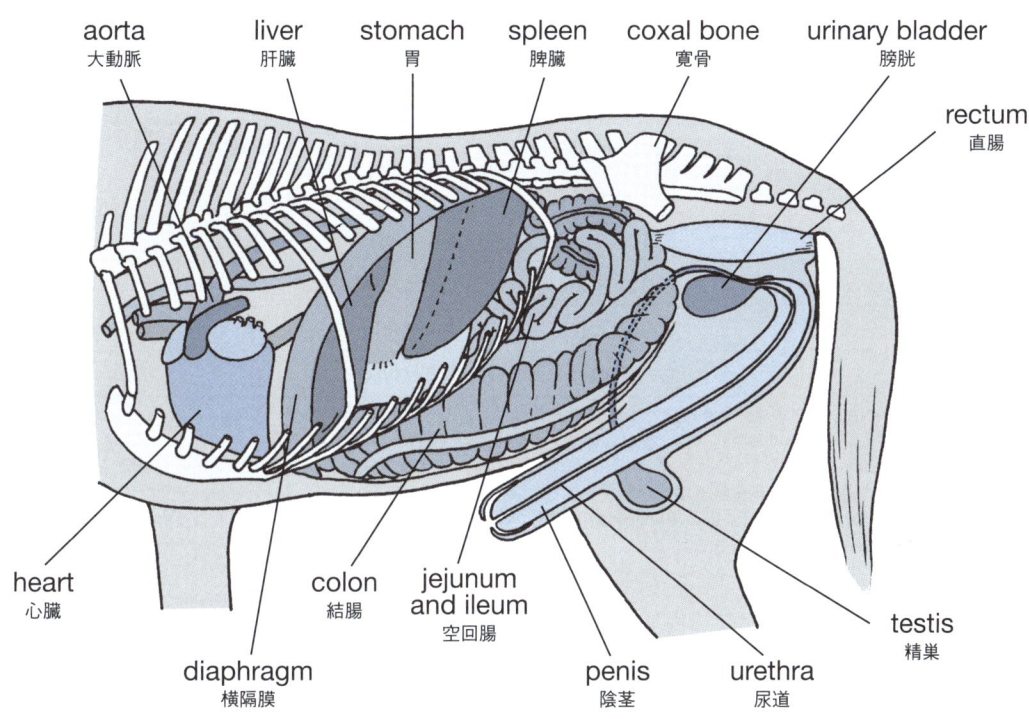

7 英語とラテン語の関係 Latin and English

臓器を表す英語形容詞は、ラテン語名詞とよく似ている。

　下の2つの表は、臓器の英語名詞と英語形容詞を並べたものである。上の表には英語の名詞と形容詞の語幹が異なる主なものを、下の表には同じものを並べた。
　英語の形容詞とラテン語名詞とを比較すると、互いによく似ていることが分かる。

英語形容詞とラテン語が類似している臓器

英語名詞	英語形容詞	ラテン語名詞	和訳
heart	cardiac	cor	心臓
lung	pulmonary	pulmo	肺
tooth(sg.), teeth(pl.)	dental	dens(sg.), dentes(pl.)	歯
tongue	lingual	lingua	舌
stomach	gastric	ventriculus*1	胃
liver	hepatic	hepar	肝臓
gall bladder, cholecyst	cystic*2	vesica fellea	胆嚢
kidney	renal	ren	腎臓
urinary bladder	cystic*2	vesica urinaria	膀胱
brain	cerebral*3	encephalon	脳
cerebrum	cerebral*3	cerebrum	大脳

英語名詞と形容詞、ラテン語が類似している臓器名

英語名詞	英語形容詞	ラテン語名詞	和訳
right ventricle*1	right ventricular	ventriculus dexter	右心室
left ventricle*1	left ventricular	ventriculus sinister	左心室
spleen	splenic	lien	脾臓
pancreas	pancreatic	pancreas	膵臓
testis	testicular	testis	精巣
ovary	ovarian	ovarium	卵巣

*1 ventricle という英語から連想するのは「心室」であろう。しかし、ラテン語の ventriculus には、心室という意味の他に「胃」「脳室」という意味がある。共に「袋状の構造」という意味から付いた名前である。
*2 cyst（袋）は訳し方がいくつかあるややこしい単語である。液体や気体を入れる袋状のものを指し、胆嚢、膀胱の両者を意味することがある。また、組織学では「嚢胞」という意味でも使われる。
*3 「脳」を意味する brain とラテン語の encephalon は全く違うスペルだが、形容詞の cerebral は「大脳の」と意味が重なるため、共に同じスペルが用いられる。

(sg.)：単数形（singular） (pl.)：複数形（plural）

体の部位を表す英語名詞とラテン語名詞を見比べてみよう。これもまた、英語名詞の大部分はラテン語名詞と異なるが、英語形容詞はラテン語名詞と大変よく似ている。

体の各各部位　Body Regions 🔊

英語名詞	英語形容詞	ラテン語名詞	和訳
head	capital	caput	頭部
horn	horn	cornu	角
ear	aural, otic	auris	耳
eye	ocular, optic	oculus	目
nose	nasal, rhinal	nasus	鼻
mouth	oral	os	口
cheek	buccal	bucca	頬
neck	cervical	collum, cervix	頚
trunk	truncal	truncus	体幹
back	dorsal	dorsum	背
loin, waist	lumbar	lumbus	腰
thorax	thoracic	thorax	胸郭
chest	pectoral	pectus	胸
rib	costal	costa	肋骨
abdomen, belly	abdominal	abdomen	腹*
navel, umbilicus	umbilical	umbilicus	臍
hip	coxal	coxa	寛骨部
gluteal region, cluneal region, buttocks, breech	gluteal, cluneal	nates, clunes	臀部
tail	caudal	cauda	尾
forelimb, thoracic limb	—	membrum thoracicum	前肢
upper arm	brachial	brachium	上腕
elbow	cubital	cubitus	肘
forearm	antebrachial	antebrachium	前腕
wrist	carpal	carpus	手根
hand	manual	manus	手
finger	digital	digiti manus	手の指
hindlimb, pelvic limb	—	membrum pelvinum	後肢
thigh	femoral	femur	大腿
knee, stifle	genicular	genu	膝
lower leg	crural	crus	下腿
ankle	tarsal	tarus	足根
foot	pedal	pes	足
toe	—	digiti pedis	足指（趾）

* 腹部のことを口語では stomach ということがある。この時は、臓器を意味しない。

8 人と動物の英語名
English Terminology of Humans and Animals

動物名の英語表記と、ラテン語由来の形容詞について知っておこう。また、形容詞の終わりのスペルが、すべて"-ine"であることにも注目しよう。

動物名の英語とラテン語の比較

英語名詞	英語形容詞	学名（ラテン語）	和訳
man	human	*Homo sapiens*	人
dog	canine*	*Canis familiaris*	犬
cat	feline	*Felis domestica*	猫
cow, cattle	bovine	*Bos taurus*	牛
horse	equine	*Equus caballus*	馬
pig, swine	porcine	*Sus scrofa*	豚
goat	caprine	*Capra hircus*	山羊
sheep	ovine	*Ovis aries*	羊
mouse	murine	*Mus musculus*	マウス

＊ 犬を"K9"と書くことがあるが、canine（ケーナイン）と音が同じであることからきたシャレである。

動物名のそれぞれの変化

名詞(sg.)	名詞(pl.)	オス	メス	子孫	和訳
man, human	men, humans	man (sg.), men (pl.)	woman (sg.), women (pl.)	baby, child (sg.), children (pl.)	人
dog	dogs	male dog	female dog	puppy	犬
cat	cats	male cat	(she-cat)	kitten	猫
cow, cattle *1	cows, cattle	bull, steer, *1 ox (sg.), oxen (pl.)	cow	calf (sg.), calves (pl.)	牛
horse	horses	stallion	mare	foal, *2 colt, filly	馬
pig, swine, hog *3	pigs, swine	boar	sow	piglet, pig	豚
goat	goats	male goat	female goat	kid	山羊
sheep	sheep	ram, tup	ewe	lamb	羊
mouse	mice	male mouse	female mouse	―	マウス

＊1 cow, cattle は総称。bull：去勢していないオス牛、steer：早期去勢牛、肉用、ox：去勢牛、労役用。
＊2 foal は総称。colt：オスの子馬、filly：メスの子馬。
＊3 hog：米国では成豚。また、去勢したオス豚を意味することもある。

Chapter 2

接頭辞
Prefixes

1 接頭辞 Prefixes

接頭辞は言葉の意味を補い、方向、数、形などを表す。

　以下に接頭辞を分類した表を載せた。これらを全部暗記しようというわけではないので、安心してほしい。これらの表を見ると、すでに知っている単語と、知らない単語があるだろうが、ここでは各々の単語に共通する「接頭辞」という考え方を、理解していただきたいのである。

　言葉を分解して考えるのは、時として難しい。例えば endorse という言葉がある。「(書状、小切手などに)裏書する」ということから、「是認する、支持する」という意味も持つようになった。endorse には一見、endo-（内）という接頭辞が隠れているように見えるが、実は、endo- + rse ではなく、en- + dorse なのである。en- は、「in（中）、on（上）」という意味であり、dorse (dorsum) は、「背中側、裏側」という意味である。すなわち「dorsum（裏）」の「on（上）」に書くから「endorse（裏書き）」となるのである。更に、ひとつの接頭辞が必ずしもひとつの意味だけを持つとは限らない。例えば、congenital（先天的な）の con- は「〜と共に」という意味であるが、conceal（隠す）の con- は、「完全に（completely）」という意味である。また、concede（しぶしぶ事実と認める）、concise（簡潔な）、conclude（結論づける）の con- は、「すべて完全に」の意味である。

　それでもなおかつ、接頭辞、語幹、接尾辞に分解して学習すると、丸覚えでなく理論的に考えられ、理解しやすく記憶しやすい。うまく理解できた時の喜びも、またひとしおである。来たな、exo- と ecto- は相同だな、と推測することができるようになる。

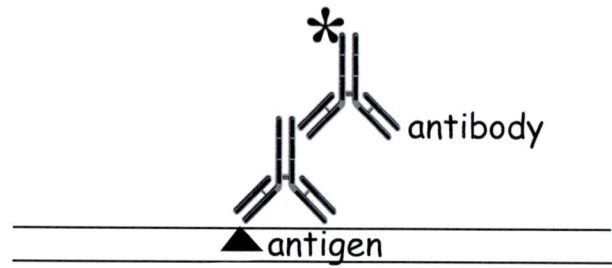

「反対、対立」を表す接頭辞

接頭辞	意味	獣医学／医学用語		一般用語
anti-	反対の	**anti**body **anti**cancer **anti**septic	抗体 抗癌の 防腐剤	anticlimax クライマックスのない。竜頭蛇尾な antismoking 禁煙の
contra-	反対の	**contra**lateral*1 **contra**ception*2	対側の 避妊	contradictory 矛盾した contrast 対照
counter-	反対の	**counter**conditioning	反対条件付け	countercharge 反論

＊1 「同側」は ipsilateral。
＊2 「妊娠、受胎」は conception。

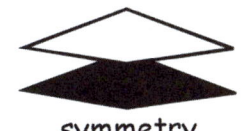
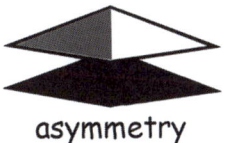

symmetry　　asymmetry

「否定」を表す接頭辞

接頭辞	意味	獣医学／医学用語		一般用語
a-, ab-	〜でない*1	**a**nemia **a**galactia **a**trophy **ab**axial **ab**lactation	貧血 乳汁分泌欠如 萎縮［症］ 反軸側の 離乳	amorphous 不定形の asexual 性別のない asymmetric 不均整の abnormal 異常な abortion 妊娠中絶
im-, in-	〜でない	**im**mortal **in**somnia*2 **in**sane	不死の 不眠症 精神錯乱の	impossible 不可能 inaccurate 不正確な inadmissible 許しがたい
non-	非、不、無	**non**invasive **non**pathogenic **non**immune	非侵襲性の 非病原性の 非免疫性の	nonfiction ノンフィクション nonsense ナンセンス
un-	〜でない	**un**conscious **un**myelinated fiber	無意識 無髄線維	unbelievable 信じがたい unreal 非現実的な

*1　a-、ab- 以下の言葉を打ち消す。
*2　in- + somnus（睡眠）。

「良し悪し」を表す接頭辞

接頭辞	意味	獣医学／医学用語		一般用語
dys-	不良、異常	**dys**lexia **dys**pepsia **dys**entery **dys**trophy	失語症 消化不良、胃弱 赤痢 ジストロフィー	dystopian 陰鬱な dysfunction 機能不全
mal-	悪い	**mal**ignant **mal**formation	悪性 奇形	maladjusted 適応障害 malady 病気、疾病
mis-	誤った	**mis**carriage **mis**diagnosis	流産 誤診	mistake 間違い mislead 誤って導く
eu-	正しい、良い	**eu**genic **eu**chromatin **eu**trophy	優生学的な 正染色質 栄養良好	euphony 快い音調 euphemism 婉曲語法 eulogy 賞賛、賛辞
ortho-	正しい	**ortho**dontics **ortho**pedics	歯科矯正学 整形外科［学］	orthodox オーソドックス

「反復」を表す接頭辞

 re

接頭辞	意味	獣医学／医学用語		一般用語
re-	再び	**re**combinant **re**activate	組み換え型 再活性化させる	react 反応する repair 修理する

「〜と共に」を表す接頭辞

接頭辞	意味	獣医学／医学用語		一般用語
co-, com-, con-	共に	**com**missure **com**ponent **con**genital **co**efficient **co**localization	交連 構成成分 先天的な 係数、率 共存	compose 構成する combine 結合する contribute 寄与する cooperate 協力する coordinate 調和させる
sym-, syn-	〜と共に	**sym**physis **sym**ptom **syn**graft **syn**apse **syn**drome	線維軟骨結合 症状 同種（同族）移植 シナプス 症候群	sympathy 同情 symmetry 対称 synchronize 同時に起こる synopsis 大意、あらすじ （劇、論文など）

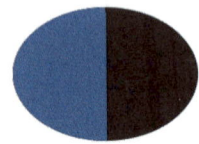

hemiは「正確に半分」、semi-とdemi-は「準、完全ではない」というイメージ。

「半分、準」を表す接頭辞

接頭辞	意味	獣医学／医学用語		一般用語
hemi-	半分	**hemi**section cerebral **hemi**sphere	半側切断 大脳半球	hemisphere 半球
semi-	半分、準ずる	**semi**tendinous muscle **semi**membranous muscle	半腱様筋 半膜様筋	semiannual 半年ごとの semiformal 準公式の
demi-	半分	**demi**lune	半月（体）*	demigod 半神半人

＊ 関節半月は articular meniscus。

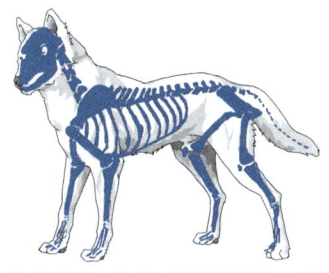

endoskeleton（内骨格）
：体の芯に硬い骨格を持ち、骨格の周りは軟らかい。

exoskeleton（外骨格）
：体の周囲に硬い骨格を持ち、体内は軟らかい。

「内外」を表す接頭辞―その1

接頭辞	意味	獣医学／医学用語		一般用語
endo-	内	**endo**crine **endo**cytosis **endo**toxin	内分泌 エンドサイトーシス 内毒素	endogamy 同族結婚
exo-, ex-	外	**exo**crine **exo**cytosis	外分泌 開口分泌	exogamy 族外結婚 exit 出口

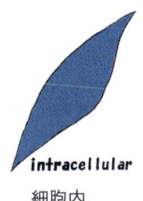
intracellular
細胞内

extracellular
細胞外

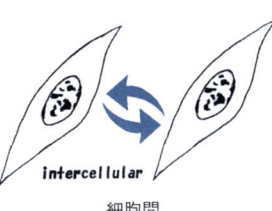
intercellular
細胞間

「内外」を表す接頭辞―その2 *

接頭辞	意味	獣医学／医学用語		一般用語
intra-	内部に	**intra**venous injection **intra**cellular	静脈注射 細胞内	intramural 学内、城壁内、建物内
extra-	外部に	**extra**cellular **extra**uterine pregnancy	細胞外 子宮外妊娠	extramural 構外、郭外 extramarital 婚外、夫婦外の
inter-	間に	**inter**calated disc	介在板	interact 相互作用

＊ 意味は endo-, exo- と同様だが、固有の語幹を持つものである。例えば、細胞内は intracellular であり、endocellular ではない。どの語幹が endo-、もしくは intra- と結合しうるのかは、個々の言葉によって習慣的に決まっている。

「近傍」を表す接頭辞

接頭辞	意味	獣医学／医学用語		一般用語
para-	近傍の、 並んで	**para**crine **para**site[*1] **para**lyze	傍分泌 寄生虫 麻痺させる[*2]	paragraph 節、段落 parallel 平行の paradox パラドックス

*1　parasite = para- + site（sitos：食物）。「横にいて食べる者」の意より。
*2　ギリシャ語で「片側の筋肉のゆるみ」の意より。

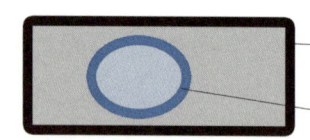
perimeter
peri- はそのものの形によらない。四角でも丸でも、その周囲のことを言う。
circumference
circum- が付くのは、その形が丸い (circle) 時。

「周囲」を表す接頭辞

接頭辞	意味	獣医学／医学用語		一般用語
peri-*	周りの、周囲の	**peri**cardium **peri**osteum	心膜 骨膜	periscope 潜望鏡 periphery 辺縁、末梢
circum-	周囲を取り巻く	**circum**vascular **circum**flex	血管周囲の 回旋の	circumstance 状況 circumference 円周

* peri- は位置の周囲だけではなく、時間的にも「一付近」として使われる。例えば、周産期は perinatal period。

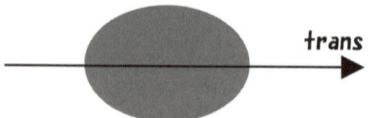

「貫通」を表す接頭辞

接頭辞	意味	獣医学／医学用語		一般用語
trans-	横切って	**trans**membrane **trans**fusion	膜を貫通した 輸液	translucent 半透明の transfer 移動
dia-	通して	**dia**phragm **dia**rrhea	横隔膜 下痢	diagram 図
per-	通して、介して	**per**cutaneous **per**oral	経皮の 経口の	perfusion 灌流

「～ならしめる、～をやめる」を表す接頭辞

接頭辞	意味	獣医学／医学用語		一般用語
em-, en-	～の中に、～ならしめる	**em**pathy **en**demic*2	感情移入*1 風土病の	embark 積み込む enter 入る encode 暗号に変える
de-	引き離す、下げる	**de**fibrillation **de**toxication **de**capitate	除細動 解毒 断頭	decode 暗号を解く
dis-	離す	**dis**sect **dis**infect	切り開く 消毒する	disabled 身体障害の disconnect 切り離す

*1 「共感」は sympathy。
*2 demos はギリシャ語で「人々」。

```
pre-op                    post-op
(preoperative)  operation  (postoperative)
    術前          手術          術後              time
```

「前後」を表す接頭辞

接頭辞	意味	獣医学／医学用語		一般用語
ante-	前方の	**ante**brachial **ante**rior	前肢の 前方の	AM (ante meridian) 午前 antecedent 先行する
pre-, pro-	前 (before)	**pre**natal **pre**cancerous **pro**gnosis* **pro**state	出生前（胎児期）の 前癌状態の 予後 前立腺	precursor 先駆者 progenitor 先祖
post-	後 (after)	**post**erior **post**natal **post**mortem	後方の 出生後の 死後の	PM (post meridian) 午後 postcoital 性交後の posthumous 死後の

* -gnosis は「知る」。診断は diagnosis。

「中、後」を表す接頭辞

接頭辞	意味	獣医学／医学用語		一般用語
mes-	中央の	**mes**encephalon **mes**oderm **mes**enchyme **mes**angium	中脳 中胚葉 間葉 腎臓の糸球体間質	mesopotamia メソポタミア （チグリス川、ユーフラテス 川の間の意） mesozoic 中生代
meta-*	の後	**met**anephros **met**atarsus **met**encephalon	後腎 中足 後脳	metamorphosis 変態、変形

* 母音および h の前では met-。

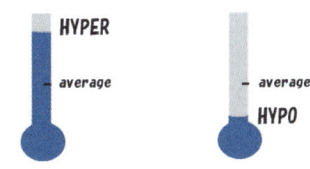

「上下」を表す接頭辞

接頭辞	意味	獣医学／医学用語		一般用語
hyper-	上に、 過度に	**hyper**tension **hyper**acidity	高血圧 胃酸過多症	hyperactive 極度に活動的な hypersensitive 過敏な
hypo-	下に、 過少に	**hypo**tension **hypo**thalamus	低血圧 視床下部*	hypoactive 活動が弱すぎる hypothesis 仮定、前提

* 視床上部は hyperthalamus ではなく epithalamus。

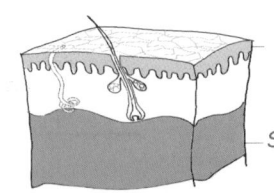

「広範囲への広がりと上下」を表す接頭辞

接頭辞	意味	獣医学/医学用語		一般用語
pan-	汎	**pan**demic **pan**angiitis **pan**neuritis	汎流行性の 汎血管炎 汎神経炎	Pan-American 全米（汎米）の
epi-	上	**epi**dermis **epi**thelium **epi**demic	表皮 上皮 流行性、伝染性の	epithet 形容辞 epilogue エピローグ、結びの言葉
sub-	下	**sub**cutaneous injection **sub**arachnoid hemorrhage	皮下注射 くも膜下出血	submarine 潜水艦 substandard 標準以下の

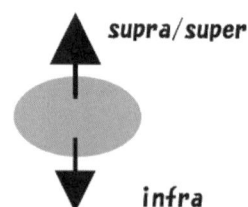

「位置的な上下」を表す接頭辞

接頭辞	意味	獣医学/医学用語		一般用語
super-	上、 超越した	**super**acute **super**ficial	極急性 表面、浅	superimpose 画像を重ね合わせる superiority complex 優越感
supra-	上	**supra**spinatus muscle **supra**orbital nerve	棘上筋 眼窩上神経	supranational 超国家的 the Supreme Court 最高裁判所
infra-	下	**infra**spinatus muscle **infra**orbital nerve	棘下筋 眼窩下神経	infrared 赤外線 inferiority complex 劣等感

「相同性と相違性」を表す接頭辞

接頭辞	意味	獣医学/医学用語		一般用語
hetero-	異なる	**hetero**geneous **hetero**chromatin	不均質の 異染色質	heterosexual 異性愛の heterodoxy 異端
homo-, homeo-	同じ	**homo**geneous **homeo**stasis	均質の ホメオスタシス （恒常性）	homosexual 同性愛の homology 相同性

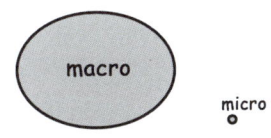

「大小」を表す接頭辞

接頭辞	意味	獣医学／医学用語		一般用語
mega-	巨大な	**mega**colon **mega**karyocyte	メガコロン（巨大結腸） 巨核球	megaton 100万トン
macro-	大きい	**macro**phage **macro**molecule	大食細胞 高分子	macrocosm 大宇宙
micro-	微小な	**micro**scopy **micro**biology	顕微鏡 微生物学	micron ミクロン（1000分の1ミリ）
parvo-	小さい	**parvo**cellular **parvo**virus	小細胞体の パルボウイルス	―

「時間」を表す接頭辞

接頭辞	意味	獣医学／医学用語		一般用語
chrono-*	時間	**chrono**graph	時間に沿った測定記録装置	chronological 年代順の

* ギリシャ語の choronos（時間）より。ギリシャ神話の時間の神の名でもある。

「色」を表す接頭辞

接頭辞	意味	獣医学／医学用語		一般用語
alb-*1	白	**alb**ino **alb**umin **alb**umen linea **alb**a	アルビノ、白子 アルブミン 卵白 白線*2	album アルバム（ラテン語「白い書字板」の意）
xanth(o)-*3	黄色	**xanth**ine **xanth**emia **xanth**opsia **xanth**oma	キサンチン 黄色血症 黄視症 黄色腫	―
cyano-*4	濃紺	**cyano**sis **Cyano**bacteria	チアノーゼ シアノバクテリア、青緑色細菌門	cyan 青緑色の cyanide シアン化物
glauco-*5	青緑色	**glauco**ma	緑内障	―
melan(o)-, nigra*6	黒	**melan**in **melan**oma substantia **nigra**	メラニン メラノーマ、黒色腫 黒質（中脳）	Melanesia メラネシア（黒い島の意） niger 黒人の蔑称

*1 ラテン語 albus より。
*2 腹部正中の結合組織。
*3 ギリシャ語 xanthos より。
*4 ギリシャ語 kyanōsis より。
*5 ギリシャ語 glaukos より。
*6 ギリシャ語 melas、ラテン語 niger より。

「数」を表す接頭辞

接頭辞		意味	獣医学／医学用語		一般用語
1	uni-	1つの(L)	**uni**lateral	一側性の	universe 宇宙 uniform 同型の、制服
	mono-	1つの(G)	**mono**clonal **mono**mer	モノクローナル 単量体	monopoly 独占、専売 monotone 一本調子の
2	bi-	2つの(L)	**bi**ceps **bi**polar **bi**section	二頭筋 双極性 両断	biannual 年2回の bilingual 2か国語に堪能な bisexual 両性の
	di-	2つの(G)	**di**mer **di**amine	二量体 ジアミン	dichotomy 2分、両分
3	tri-	3つの(L,G)	**tri**ceps brachii hepatic **tri**ad	上腕三頭筋 肝三つ組	trio トリオ、3つ組 trilingual 3か国語に堪能な
4	quadri-, quadro-	4つの(L)	**quadri**ceps femoris **quadri**plegia	大腿四頭筋 四肢麻痺	quartet カルテット、4つ組 quadrilateral 四辺形の
	tetra-	4つの(G)	**tetra**pod **tetra**cycline	四足獣 テトラサイクリン	tetragon 四角形 tetrahedron 四面体
multi-		多数の(L)	**multi**vesicular body	多胞小体	multiple 多数の
poly-		多数の(G)	**poly**clonal **poly**mer	ポリクローナル ポリマー重合体	polygamist 一夫多妻者 polyglot 数か国語を話す

(L) ラテン語起源、(G) ギリシャ語起源。

 **40**

　検疫（quarantine）という英語はラテン語の40（quarantina）に由来する。ペストが大流行した中世ヨーロッパにおいて、1377年、ヴェネチアはペストの侵入をおそれ、船舶を40（quarantina）日（最初は30日、後に短すぎるとして10日延長し40日にした）停泊させて様子を見た。これが検疫（quarantine）の始まりである。

問1．次の接頭辞と意味を正しく結びなさい。

1. ①上 ・　　　・ a) ante-
   ②下 ・　　　・ b) hypo-
   ③前 ・　　　・ c) hyper-
   ④後 ・　　　・ d) post-

2. ①反 ・　　　・ a) hetero-
   ②非 ・　　　・ b) anti-
   ③同 ・　　　・ c) homo-
   ④異 ・　　　・ d) un-

3. ①内 ・　　　・ a) exo-
   ②外 ・　　　・ b) endo-
   ③上 ・　　　・ c) inferior-
   ④下 ・　　　・ d) superior-

4. ①汎 ・　　　・ a) sub-
   ②上 ・　　　・ b) pan-
   ③下 ・　　　・ c) super-

5. ①内 ・　　　・ a) inter-
   ②外 ・　　　・ b) extra-
   ③間 ・　　　・ c) intra-

問2．次の用語の左右を正しく結びなさい。

1. ①三角形 ・　　　・ a) polygonal
   ②四角形 ・　　　・ b) triangle
   ③多角形 ・　　　・ c) quadrangle, tetragon

2. ①単量体 ・　　　・ a) polymer
   ②二量体 ・　　　・ b) monomer
   ③三量体 ・　　　・ c) dimer
   ④多量体 ・　　　・ d) trimer

3. ①直腸内の ・　　　・ a) interspace
   ②間腔 ・　　　・ b) intrarectal
   ③細胞外基質 ・　　　・ c) extracellular matrix (ECM)
   ④間質性の ・　　　・ d) extrapyramidal tract
   ⑤錐体外路 ・　　　・ e) interstitial

4. ①高血糖　　　　　・　　　・　a) hypertension
　　②低血糖　　　　　・　　　・　b) hyperglycemia
　　③高血圧　　　　　・　　　・　c) hypotension
　　④低血圧　　　　　・　　　・　d) hypoglycemia
　　⑤あら捜しする　　・　　　・　e) hypocritical
　　⑥偽善的な　　　　・　　　・　f) hypercritical

問3．次の言葉を接頭辞［ab-／im-／in-／non-／dys-／un-］のいずれかを使って反対の意味に変えなさい。

① axial → _____
② active → _____
③ invasive → _____
④ mortal → _____
⑤ function → _____
⑥ conscious → _____
⑦ digestion → _____

問4．接頭辞と語幹を組み合わせて、言葉を作りなさい。

例	a- + emia	anemia	貧血
①	dys- + uria		排尿障害
②	poly- + uria		多尿
③	olig- + uria		乏尿
④	olig- + dendrocyte		稀突起膠細胞
⑤	anti- + septic		防腐性の
⑥	para- + thyroid		上皮小体
⑦	supra- + renal		副腎の
⑧	sub- + cellular		細胞小器官レベルの

Exercises 1 の答え

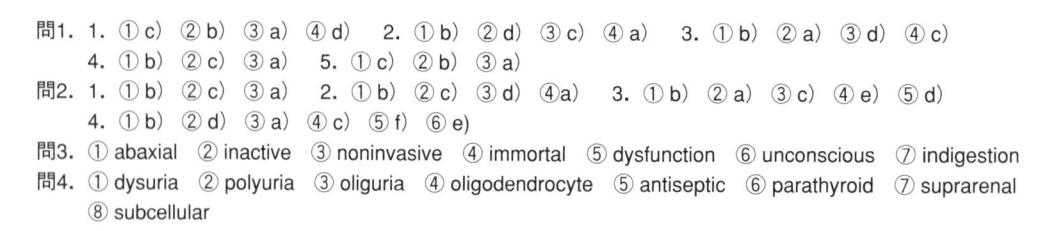

問1．1. ①c) ②b) ③a) ④d)　2. ①b) ②d) ③c) ④a)　3. ①b) ②a) ③d) ④c)
　　4. ①b) ②c) ③a)　5. ①c) ②b) ③a)
問2．1. ①b) ②c) ③a)　2. ①b) ②c) ③d) ④a)　3. ①b) ②a) ③c) ④e) ⑤d)
　　4. ①b) ②d) ③a) ④c) ⑤f) ⑥e)
問3．① abaxial　② inactive　③ noninvasive　④ immortal　⑤ dysfunction　⑥ unconscious　⑦ indigestion
問4．① dysuria　② polyuria　③ oliguria　④ oligodendrocyte　⑤ antiseptic　⑥ parathyroid　⑦ suprarenal
　　⑧ subcellular

Chapter 3

接尾辞
Suffixes

1 接尾辞 Suffixes

接尾辞は言葉の意味を補い、症状、術式などを表す。

　例えば -tomy は「切る」という意味を持つ接尾辞である。解剖学の anatomy はここに由来する。開腹術や切開術には、-tomy という接尾辞が多く用いられる。（anatomy の ana- は、「上へ、再び、後ろへ」を意味する。analysis（分析する）の ana- と同様である）

接尾辞は語尾変化する。

　anatomist（解剖学者）という言葉を分解して考えると、-tomy という接尾辞に、「〜する人」を意味する接尾辞 -ist が加わった言葉である。
　このように、名詞から名詞へというだけでなく、接尾辞は文に合わせて語尾変化する。例えば「解剖学的な」という形容詞は anatomical、「解剖学的に」という副詞は anatomically というようにである。

　また、接尾辞が変わると意味も変わる。例えば、-scope（鏡、計、器）という接尾辞は、「眼で見て調べる機器」という意味を含み、microscope は、micro-（小さいことを意味する接頭辞）、すなわち細かいものを「見て調べる機器」だから「顕微鏡」を意味する。この microscope は1台、2台、と数えることのできる機器そのものを指す。

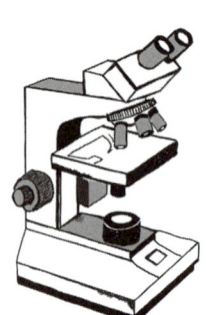

「見て調べる機器とそれを用いた検査法」を表す接尾辞

接尾辞	意味	獣医学／医学用語		一般用語
-scope	鏡、計、器	microscope bronchoscope gastroscope stethoscope vaginoscope (L) *, colposcope (G) *	顕微鏡 気管支鏡 胃鏡 聴診器 膣鏡	telescope 望遠鏡 kaleidoscope 万華鏡
-scopy	観察［法］ 検査［法］	microscopy bronchoscopy gastroscopy stethoscopy vaginoscopy (L) *, colposcopy (G) *	顕微観察［法］ 気管支鏡検査［法］ 胃内視鏡検査［法］ 聴診［法］ 膣鏡検査［法］	horoscopy 占星術

＊　vagin- も colpo- も「膣」という意味で、前者はラテン語（L）起源、後者はギリシャ語（G）起源である。

この接尾辞が -scopy（観察[法]、検査[法]）に変わると、これは機器を指すのではなく microscopy「顕微鏡を用いて行う検査」を指し、もはやひとつ、ふたつとは数えられない。

このように、microscope から microscopy へとアルファベット1文字で、意味は「機器」から「検査」へと変化するのである。

「計測」を表す接尾辞*

接尾辞	意味	獣医学／医学用語		一般用語
-meter	計、計測器、測定器	densito**meter** thermo**meter** spectro**meter** cyto**meter** audio**meter**	密度計、比重計 温度計、体温計 分光計 細胞計算器 聴力計	barometer 気圧計、晴雨計
-metry	測定[法]、計測[法]、定量[法]	densito**metry** thermo**metry** spectro**metry** cyto**metry** audio**metry**	密度（濃度）計測[法] 温度測定、検温 分光（光度）法 生体計測 聴力検査[法]	geometry 幾何学 uranometry 天体図

* -meter は測定機器そのものを、-metry はそれを使った検査法を示す。gas meter や parking meter は2語から成るので、これらの meter は接尾辞ではない。

「画像処理」にかかわる接尾辞*1

接尾辞	意味	獣医学／医学用語		一般用語
-gram	図、像、描かれた、書かれたもの	electrocardio**gram**（ECG） electroencephalo**gram**（EEG） scinti**gram** ultrasono**gram**	心電図*2 脳波図 シンチグラム 超音波検査図	program プログラム phonogram 表音文字 tachogram タコグラム、速度図
-graph	計、器、装置	electrocardio**graph** electroencephalo**graph** oscillo**graph** thermo**graph** tomo**graph**	心電計 脳波計 オシログラフ 温度記録計 断層撮影装置	photograph 写真 pictograph 絵文字

*1 -graph と -graphy の関係も -scope と -scopy の関係と同様で、-graphy は、「記録法、撮影法、造影法、測定法、学、論」を意味する。
*2 EKG（electrocardiogram）ともいう。ドイツ語の Electrokardiogramm より。

 piriform    lentiform    fungiform    punctiform

「形状」を表す接尾辞

接尾辞	意味	獣医学／医学用語		一般用語
-form	状、形、様	piri**form**, pyri**form** lenti**form** fili**form** fungi**form** puncti**form**	梨状の レンズ状の 糸状の 茸状の 点状の	cruci**form** 十字架の
-oid	類、様	alkal**oid** arachn**oid** myel**oid** sesam**oid** lip**oid**	アルカロイド、植物塩基 くも膜の 骨髄（性）の ゴマ状、種子骨の 類脂質	cub**oid** 立方形 spher**oid** 楕円の human**oid** ヒトそっくりの、ヒト型ロボット

「新生、形成」を意味する接尾辞＊

接尾辞	意味	獣医学／医学用語		一般用語
-genesis	発生、形成	angio**genesis** carcino**genesis** neuro**genesis** spermato**genesis** oo**genesis**	血管新生 発癌、癌発生 神経発生 精子形成 卵子形成	poly**genesis** 多元発生説 meta**genesis** 世代交代
-poiesis	形成、生成	angio**poiesis** erythro**poiesis** hemato**poiesis** myelo**poiesis** thrombo**poiesis**	血管形成［新生］ 赤血球造血 造血 骨髄形成 血餅形成	auto**poiesis** 自動生成

＊ -genesis は、「創始、新生、発生、（新しく作る）」、-poiesis は「生産する（making）」を意味する。

 Column 織物の上手な娘

arachnoid（くも膜）はギリシャ語の arachne（くも）＋ -oid に由来し、くもの巣のように入り組んだ軟らかい膜構造である。ギリシャ神話の昔、織物の上手な Arachne（アラクネ）という娘がいた。彼女は、「自分の織物の腕は、戦と工芸の女神アテネに勝るとも劣らない」と誇ったため、女神の怒りに触れくもにされてしまう。くもという言葉は、可愛そうな彼女の名前に由来する。

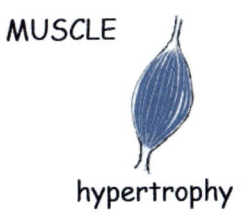
hypertrophy

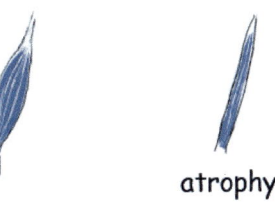
atrophy

MUSCLE

「栄養」を意味する接尾辞

接尾辞	意味	獣医学／医学用語	
-trophia, -trophy	栄養[症]	atrophia, atrophy*	萎縮[症]、無栄養[症]
		hypertrophia, hypertrophy	肥大、栄養過度
		dystrophia, dystrophy	ジストロフィー、異栄養、栄養失調
		neurotrophy	神経栄養
		lipotrophy	脂肪増多症

* 打ち消しの接頭辞 a- が付くと、萎縮症（名詞）を表す。

「分野、主義、人、精神状態」を表す接尾辞

接尾辞	意味	獣医学／医学用語		一般用語
-ics	学、法、術	statistics	統計学	ethics 倫理学
		proteomics	プロテオミクス	gymnastics 体操
		morphometrics	形態計測[法]	
		genetics	遺伝学	
		biodynamics	生体力学	
-ism	主義、症、状態	autism	自閉症	individualism 個人主義
		hyperthyroidism	甲状腺機能亢進症	optimism 楽観主義
		hypoinsulinism	低インスリン症	
		phobism	恐怖[症]状態	
		rheumatism	リウマチ	
-ist	士、者、学者	anesthesiologist	麻酔医	individualist 個人主義者
		microbiologist	微生物学者	optimist 楽天家
		geneticist	遺伝学者	

「病的な好き、嫌い」を表す接尾辞

arachnophobia

接尾辞	意味	獣医学／医学用語	
-philia	（病的）愛好［症］	eosinophilia hemophilia pedophilia	好酸球増多症 血友病 小児性愛
-phobia	（病的）恐怖［症］	hydrophobia anthropophobia arachnophobia	恐水症（狂犬病） 対人恐怖症 クモ恐怖［症］

pyromania

「狂うほどの好き」を表す接尾辞

接尾辞	意味	獣医学／医学用語	
-mania	癖、狂	dipsomania egomania pyromania nymphomania monomania pseudomania hypomania	飲酒癖、渇酒癖 病的な（極端な）自己中心癖 放火癖、放火狂 色情狂の女性 偏執狂 病的虚言癖 軽躁病

「殺す」を意味する接尾辞

接尾辞	意味	獣医学／医学用語		一般用語
-cide	殺し	pesticide*1 rodenticide*2 insecticide*3 bactericide*4 germicide*5	殺虫剤 殺鼠薬 殺虫剤 殺菌薬 殺菌剤	homicide 人殺し suicide 自殺

*1 pest（害虫）＋ -cide。
*2 rodent（げっ歯類）＋ -cide。
*3 insect（昆虫）＋ -cide。
*4 bacteria（細菌）＋ -cide。
*5 germ（病原菌）＋ -cide。

「疫病」を表す接尾辞*

接尾辞	意味	獣医学／医学用語		一般用語
-ia	態、病態	hern**ia** malar**ia** pneumon**ia** dysur**ia** dystoc**ia**	ヘルニア マラリア 肺炎 排尿困難 難産	hysteria ヒステリー
-sis	症	cyano**sis** adeno**sis** sclero**sis** endometrio**sis** keto**sis** zoono**sis**	チアノーゼ 腺症 硬化症 子宮内膜症 ケトン症、ケトーシス 人獣共通感染症	halitosis（bad breath）口臭

* 炎症を表す -itis については p.42〜、腫瘍を表す -oma, -sarcoma については腫瘍学の p.158〜、結石については胆嚢と胆管のp.117と泌尿器のp.124-125を参照のこと。

「疾患」を表す接尾辞

接尾辞	意味	獣医学／医学用語		一般用語
-pathia -pathy	疾患、 疾病	angio**pathy** myo**pathy** neuro**pathy** nephro**pathy** retino**pathy**	脈管障害 筋障害 神経障害 腎障害 網膜症	sympathy 同情 empathy 共感、感情移入

「溶解」を表す接尾辞

接尾辞	意味	獣医学／医学用語		一般用語
-lysis	離開、 融解、 溶解、 分解	para**lysis** auto**lysis** osteo**lysis** proteo**lysis** hemo**lysis** thrombo**lysis**	（完全）麻痺、中風 自己溶解 骨溶解 蛋白質分解 溶血 血栓溶解	analysis 分析、分解 hydrolysis 加水分解

「漏れること」を表す接尾辞

接尾辞	意味	獣医学／医学用語	
-rrhea -rrhoea	漏	dia**rrhea** ameno**rrhea** polymeno**rrhea**, epimeno**rrhea** cholo**rrhea** galacto**rrhea**, acto**rrhea** pyo**rrhea**	下痢 無月経 頻発月経 胆汁過剰分泌 乳汁漏出［症］ 膿漏［症］

「破裂」を表す接尾辞

接尾辞	意味	獣医学／医学用語	
-rrhexis	破裂	amnio**rrhexis**	破水、羊膜破裂
		arterio**rrhexis**	動脈破裂
		cardio**rrhexis**	心臓破裂（心壁破裂）
		kerato**rrhexis**	角膜破裂
		myo**rrhexis**	筋断裂

「ヘルニア、瘤」を表す接尾辞

接尾辞	意味	獣医学／医学用語	
-cele	腫脹、ヘルニア、瘤	amnio**cele**	臍帯ヘルニア
		appendico**cele**	虫垂ヘルニア
		arthro**cele**	関節瘤
		entero**cele**	脱腸、腸ヘルニア
		hemato**cele**	血瘤、血腫

「麻痺」を表す接尾辞

接尾辞	意味	獣医学／医学用語	
-plegia	麻痺	di**plegia**	両側麻痺、対麻痺
		para**plegia**	対麻痺
		hemi**plegia**	半側麻痺、片麻痺
		neuro**plegia**	神経遮断
		quadri**plegia**, tetra**plegia**	四肢麻痺

「下垂症」を表す接尾辞

接尾辞	意味	獣医学／医学用語	
-ptosis	下垂［症］	blepharo**ptosis**	眼瞼下垂［症］
		gastro**ptosis**	胃下垂
		procto**ptosis**	肛門脱
		metro**ptosis**	子宮脱
		cysto**ptosis**	膀胱下垂［症］

「欠乏症」を表す接尾辞

接尾辞	意味	獣医学／医学用語	
-penia	欠乏［症］、減少［症］	erythrocyto**penia**, erythro**penia**	赤血球減少［症］
		leukocyto**penia**, leuko**penia**	白血球減少［症］
		neutro**penia**	好中球減少［症］
		glyco**penia**	糖欠乏［症］
		kalio**penia**	カリウム欠乏［症］

「切開」を表す接尾辞

接尾辞	意味	獣医学／医学用語	
-tomy*	切開[術]、切離[術]	ana**tomy**	解剖学
		hystero**tomy**	子宮切開[術]
		masto**tomy**	乳房切開[術]
		ovario**tomy**	卵巣切開[術]
		hymeno**tomy**	処女膜切開[術]
		uretero**tomy**	尿管切開[術]

* tomy はギリシャ語の tome（切ること）に由来する。ミクロトームの tome もこれである。否定の接頭辞 a- を付けて atom となると、これ以上切ることができないもの、つまり原子となる。

「摘出術」を表す接尾辞

接尾辞	意味	獣医学／医学用語	
-ectomy*	切除[術]、摘出[術]	vas**ectomy**	精管切除[術]
		ovariohyster**ectomy**	卵巣子宮摘出術
		mast**ectomy**	乳房切除[術]
		hymen**ectomy**	処女膜切除[術]
		ureter**ectomy**	尿管切除[術]

* ectomy はギリシャ語の ektome（外に切り出すこと）に由来する。ek は exo- や ecto- と同様「外」を意味する。ただその場で切るだけが -tomy 、切ったものを外に取り出すのが -ectomy である。

「形成」を表す接尾辞

接尾辞	意味	獣医学／医学用語	
-plasia	形成	hyper**plasia**	過形成
		hypo**plasia**	低形成、形成不全
		dys**plasia**, allo**plasia**	異形成
		ana**plasia***	退形成

* ana- は「上へ、後ろへ、再び」を意味する。

「形成術」を表す接尾辞

接尾辞	意味	獣医学／医学用語	
-plasty	形成[術]	mamma**plasty**, mammo**plasty**	乳房形成[術]
		angio**plasty**	血管形成[術]
		palato**plasty**	口蓋形成[術]
		vagino**plasty**	膣形成[術]
		rhino**plasty**	鼻形成[術]*

* 口語で鼻の美容整形は nose job という。例：She got a nose job.

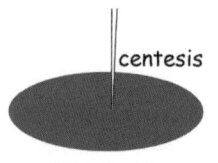

「穿刺」を表す接尾辞

接尾辞	意味	獣医学／医学用語	
-centesis	穿刺	amnio**centesis**	羊水穿刺［法］
		abdomino**centesis**	腹腔穿刺［法］
		peritoneo**centesis**	腹膜穿刺［法］
		para**centesis**	穿刺術［法］
		pleuro**centesis**	胸膜穿刺［法］

「縫合術」を表す接尾辞

接尾辞	意味	獣医学／医学用語	
-rrhaphy	縫合［術］	angio**rrhaphy**	血管縫合［術］、脈管縫合［術］
		entero**rrhaphy**	腸縫合［術］
		vaginoperineo**rrhapy**	腟会陰縫合［術］*
		achillo**rrhaphy**	アキレス腱縫合［術］
		palato**rrhapy**	口蓋縫合［術］

＊　腟と会陰の裂創修復。

「開口、吻合、造瘻術」を表す接尾辞

接尾辞	意味	獣医学／医学用語	
-stomy	開口［術］、吻合［術］、造瘻［術］	colo**stomy**＊	人工肛門形成術
		enteroentero**stomy**	腸腸吻合［術］
		arthro**stomy**	関節切開［術］
		tracheo**stomy**	気管開口［術］、気管切開［術］
		cholangio**stomy**	胆管造瘻［術］
		salpingo**stomy**	卵管開口［術］

＊　colon（結腸）＋ -stomy。

「吻合術」を表す接尾辞

接尾辞	意味	獣医学／医学用語	
-anastomosis*1	吻合［術］	esophagogastro**anastomosis**（＝esophagogastrostomy）	食道胃吻合［術］
		entero**anastomosis**	腸管吻合［術］
		micro**anastomosis**	微小吻合［術］*2
		gastro**anastomosis**	胃胃吻合［術］*3

＊1　stoma はギリシャ語で「口」という意味。anastomosis は長いフレーズなので、語幹に付けて接尾辞として使うだけでなく、2語に分けて atriovenous anastomosis（動静脈吻合）のように使われることも多い。
＊2　顕微鏡下で行われる微小構造の吻合。
＊3　胃噴門部と前庭部の吻合。

「療法」を表す接尾辞

接尾辞	意味	獣医学／医学用語		一般用語
-therapy*	療法	chemo**therapy**	化学療法	aromatherapy アロマセラピー
		radio**therapy**	放射線治療	thalassotherapy 海水療法
		physio**therapy**	物理療法、理学療法	
		cryo**therapy**	寒冷療法	
		thermo**therapy**	温熱療法	
		sero**therapy**	血清療法	

* gene therapy，shock therapy などの2語で使われることも多い。なお、animal therapy は和製英語であり、正式には animal-assisted therapy と言う。

「固定術」を表す接尾辞*

接尾辞	意味	獣医学／医学用語	
-pexy	固定［術］	ceco**pexy**	盲腸固定［術］
		entero**pexy**	腸固定［術］
		nephro**pexy**	腎固定［術］
		pneumono**pexy**	肺固定［術］
		retino**pexy**	網膜復位［術］

* 牛の第四胃変位の固定術を abomasal pexy とも言うが、これは2語であるため接尾辞ではない。

Exercises 2

問1．cardi-（心臓）と接尾辞をつないで、次の用語を作りなさい。

① 心電図 ＿＿＿＿＿＿＿＿＿＿＿＿＿＿＿
② 心臓疾患 ＿＿＿＿＿＿＿＿＿＿＿＿＿＿＿
③ 心臓破裂 ＿＿＿＿＿＿＿＿＿＿＿＿＿＿＿
④ 心臓穿刺 ＿＿＿＿＿＿＿＿＿＿＿＿＿＿＿
⑤ 心臓麻痺 ＿＿＿＿＿＿＿＿＿＿＿＿＿＿＿

問2．angio-（血管）と接尾辞をつないで、次の用語を作りなさい。

① 血管新生 ＿＿＿＿＿＿＿＿＿＿＿＿＿＿＿
② 血管形成術 ＿＿＿＿＿＿＿＿＿＿＿＿＿＿＿
③ 血管縫合術 ＿＿＿＿＿＿＿＿＿＿＿＿＿＿＿
④ 脈管障害 ＿＿＿＿＿＿＿＿＿＿＿＿＿＿＿
⑤ 血管形成異常 ＿＿＿＿＿＿＿＿＿＿＿＿＿＿＿

問3．語幹と -lith（結石）をつないで、次の用語を作りなさい。

①腎結石 _____
②膀胱結石 _____
③尿管結石 _____
④結石症 _____

問4．下の語群から（　）に適当な用語を選びなさい。

1. 病気は、英語で（①　　　　）という。病名のなかには、この言葉をそのまま含んでいるものがある。例えばアカバネ病は英語で（②　　　　）であり、ニワトリの伝染性ファブリキウス囊病は（③　　　　）、口蹄疫は（④　　　　）である。口蹄疫は、日本語と英語では、口と蹄の順番が入れ替わるところに注目。
病名に「〜不全」が付くものがある。英語の「失敗、不足」を表す（⑤　　　　）の和訳である。例えば、心不全は（⑥　　　　）または（⑦　　　　）であり、腎不全は（⑧　　　　）、急性呼吸不全は（⑨　　　　）と言う。

   a) akabane disease　　b) disease　　c) failure
   d) infectious bursal disease　　e) renal failure
   f) foot-and-mouth-disease　　g) acute respiratory failure
   h) heart failure　　i) cardiac failure

2. 病名は、接尾辞 -sis を持つものが多い。例えば、犬糸状虫症は（①　　　　）、トキソプラズマ症は（②　　　　）、結核は（③　　　　）、豚の大腸菌症は（④　　　　）、肝硬変は（⑤　　　　）と言う。これらすべてが、-sis で終わることに注目されたい。
接尾辞 -ia も「病気」を意味する。例えば、肺炎は（⑥　　　　）であり、貧血は（⑦　　　　）である。横文字そのままを、カタカナにして使っている例もある。例えば、マラリア（⑧　　　　）やヘルニア（⑨　　　　）がそれである。

   a) toxoplasmosis　　b) colibacilosis in pig　　c) pneumonia
   d) liver cirrhosis　　e) anemia　　f) hernia　　g) malaria
   h) canine dirofilariasis　　i) tuberculosis

3. 病気になると、熱（①　　　）を発することが多い。その結果、「〜熱」という病名を持つ病気がある。例えば、アフリカ豚コレラ（②　　　）や悪性カタル熱（③　　　）などである。

変わったところでは、-pest という接尾辞、あるいは単語を持つ病気がある。例えば牛疫（④　　　）がそれである。rinder とはドイツ語で牛のことであり、pest とは、病気や病害虫などのイヤなものを指す。また、殺虫剤のことを（⑤　　　）というが、この pest も同様である。牛以外に家禽ペスト（fowl pest）や、豚ペスト（⑥　　　）がある。

a) fever   b) pesticide   c) african swine fever   d) swine pest
e) malignant catarrhal fever   f) rinderpest

問5. 下の語群から（　）に適当なものを選びなさい（重複可）。

1. 肝臓は英語で（①　　　）だが、ラテン語（学名）は（②　　　）である。ここから、英語の「肝臓の」を意味する形容詞は（③　　　）となった。この形容詞を持つ獣医学領域の病気に、例えば牛の肝膿瘍（hepatic abscesses in cattle）がある。同様に、肝硬変は肝臓を意味する英語名詞形の（④　　　）を用いて、chirrhosis of the liver と言ったり、英語形容詞形を用いて、（⑤　　　）と言ったりするが、同じ意味である。

Hepar に炎症を意味する -itis という接尾辞が付くと、（⑥　　　）、すなわち「肝炎」となる。獣医学領域では、例えば犬伝染性肝炎は英語で infectious canine hepatitis と言う。

a) hepatic   b) liver   c) hepatic chirrhosis   d) hepar   e) hepatitis

2. 腎臓は英語名詞で（①　　　）であり、ラテン語は（②　　　）である。ラテン語より派生して、「腎臓の」を表す英語形容詞は（③　　　）という。この形容詞形を用いて、腎不全は（④　　　）となる。

a) Ren   b) renal   c) kidney   d) renal failure

Exercises 2 の答え

問1. ① electrocardiogram　② cardiopathy　③ cardiorrhexis　④ cardiocentesis　⑤ cardioplegia
問2. ① angiogenesis　② angioplasty　③ angiorrhaphy　④ angiopathy　⑤ angiodysplasia
問3. ① nephrolith　② cystolith　③ ureterolith　④ lithiasis
問4. 1. ① b)　② a)　③ d)　④ f)　⑤ c)　⑥ h)　⑦ i)　⑧ e)　⑨ g)
　　　2. ① h)　② a)　③ i)　④ b)　⑤ d)　⑥ c)　⑦ e)　⑧ g)　⑨ f)　3. ① a)　② c)　③ e)　④ f)　⑤ b)　⑥ d)
問5. 1. ① b)　② d)　③ a)　④ b)　⑤ c)　⑥ e)　2. ① c)　② a)　③ b)　④ d)

2 炎症 Inflammation

Point: 炎症を表す用語には、接尾辞 -itis（－炎）を用いる。

炎症を意味する接尾辞 -itis を用いる際の語幹には、臓器の英語名詞ではなく、形容詞を用いることが多い。例えば、「胃炎」は、stomachitis ではなく、gastritis となる。

下表をじっくり見て（暗記するには及ばない）、法則性を理解しよう！

臓器とその炎症 Organs and their inflammations

（青字は例外的に英語形容詞と炎症名が対応しないもの）

臓器名		炎症名	
英語名詞	英語形容詞	-itis	和訳
heart	cardiac	carditis	心炎
nose	nasal, rhinal	rhinitis	鼻炎
larynx	laryngeal	laryngitis	喉頭炎
trachea	tracheal	tracheitis	気管炎
bronchus	bronchial	bronchitis	気管支炎
lung	pulmonary	pneumonia*1	肺炎
pleura	pleural	pleuritis	胸膜炎
dental pulp	（pulpal）	odontitis, pulpitis	歯髄炎
periodontal tissue	periodontal	periodontitis	歯周炎
tongue	lingual, glossal	glossitis	舌炎
tonsil	tonsillar, tonsillary	tonsillitis	扁桃炎
stomach	gastric	gastritis	胃炎
intestine, enteron	intestinal, enteric	enteritis	腸炎
appendix	appendicular	appendicitis	虫垂炎
liver	hepatic	hepatitis	肝炎
pancreas	pancreatic	pancreatitis	膵炎
gallbladder	cystic*2, cholecystic	cholecystitis	胆嚢炎
peritoneum	peritoneal	peritonitis	腹膜炎
kidney	renal	nephritis	腎炎
urethra	urethral	urethritis	尿道炎
（urinary）bladder	cystic*2	cystitis, urocystitis	膀胱炎
oviduct	salpingian	salpingitis	卵管炎

uterus		uterine, metrical, matricial	uter**itis**, metr**itis**	子宮炎
endometrium		endometrial	endometr**itis**	子宮内膜炎
vagina		vaginal	colp**itis**, vagin**itis**	膣炎
mammary gland		mammary	mast**itis**	乳腺炎、乳房炎
bone		osteal	oste**itis**	骨炎
cartilage		cartilaginous	chondr**itis**	軟骨炎
joint		articular	arthr**itis**	関節炎
muscle		muscular	myos**itis**	筋炎
skin		dermal, cutaneous	dermat**itis**	皮膚炎
eye		optic	ophthalm**itis**	眼炎
conjunctiva		conjunctival	conjunctiv**itis**	結膜炎
cornea		corneal	kerat**itis**	角膜炎
retina		retinal	retin**itis**	網膜炎
ear	ear	otic	ot**itis**	耳炎
	external ear	-	ot**itis** externa	外耳炎
	middle ear	-	ot**itis** media	中耳炎
	internal ear	-	ot**itis** interna	内耳炎
brain		encephalic	encephal**itis**	脳炎

＊1　pneumonitis という言い方もあるが、pneumonia の方が一般的。
＊2　cystic は、「胆嚢の」あるいは「膀胱の」の両方の意味になり得る。

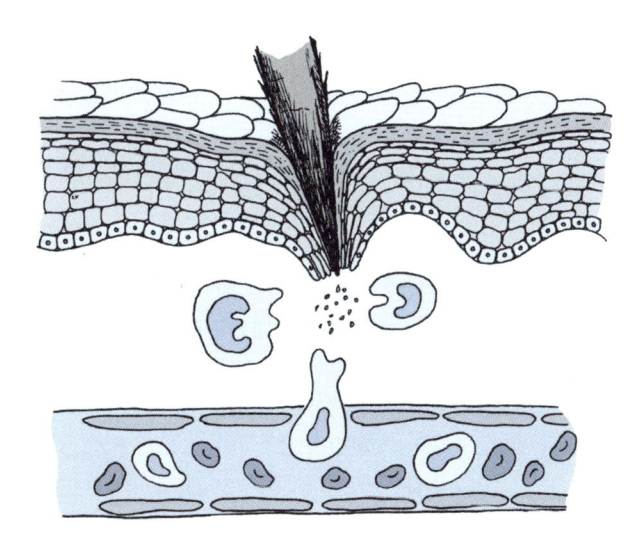

ウミ（膿）Pus

Point ウミがたまる状態を表す用語には、接頭辞 py(o)- を用いる。

炎症にウミはつきものである。ウミは細菌感染がおこった時、その免疫反応として好中球などの白血球やその細菌の死骸が、累々とたまったものである。私たちがウミを見る時、生体防御反応の熱い戦争の後を垣間見ているというわけだ。

ウミは組織から滲み出るように出てくるが、一般的に、ウミを含め、滲出物（目やに、耳だれ、膣からのおりものなど）と言うときは、discharge を用いる。

ウミ Pus

suppurat-*	suppurate	化膿する
	suppuration	化膿
py(o)-	**py**ogenesis, **py**opoiesis	化膿
	pyothorax	膿胸
	pyuria	膿尿
	hemato**py**uria	血膿尿
	pyemia	膿血症
	pyocyst	膿嚢胞
	pyometritis	化膿性子宮炎
	pyometra	子宮蓄膿症
	em**py**ema	蓄膿症
	pyorrhea	膿漏

* sup は sub-（下）、pur は pus（膿）より。

Column　光は東方より　〜ギリシャ語とラテン語〜

中近東で栄えた文明は西方、エーゲ海を内海とする現在のトルコ、ギリシャへと広がった。ギリシャ時代に医学は体系化され、紀元前4－5世紀にはヒポクラテスら専門の医師団も現れた。肺炎やコレラなど、多くの疾病名が古代ギリシャ語に由来する。

文明は更に西方へと広がり、紀元前後には強大なローマ帝国が作られ、ラテン語を主体とするローマ文化は、先立つギリシャ語をラテン語の中に取り入れていった。古代ローマを照らす光は東方、ギリシャより来たのである。やがてラテン語は日常会話で使われる言葉ではなく、文語（書き言葉）となったが、ローマ帝国がヨーロッパに勢力を伸ばすにつれて、ラテン語の読み書きはヨーロッパ知識人の不可欠な素養となっていった。解剖学の知識はラテン語をもって体系化されたのである。

英語はゲルマン語にその源流をもち、ラテン語とは起源が異なる。だが直接的、あるいはフランス語などを通して間接的に、ギリシャ語、ラテン語の影響を今も色濃く残している。

3 痛み Pain

Point: 痛みを表す用語は、接尾辞 -algia もしくは -dynia（－痛）を用いる。

望郷の念に駆られることを意味する「ノスタルジー」は、英語で nostalgia である。nostos は ギリシャ語で「家に帰る」、-algia は「痛み」、という意味であるから、「胸が痛くなるほどの望郷の念」となるのである。学術用語の「－痛」は、この -algia、もしくは同じくギリシャ語起源の -dynia を接尾辞として持つ。

下表を見ると、暗記するには至らずとも、英語の形容詞（名詞に非ず）を語幹に -algia または -dynia を接尾辞に持つ言葉がいかに多いかが分かる。

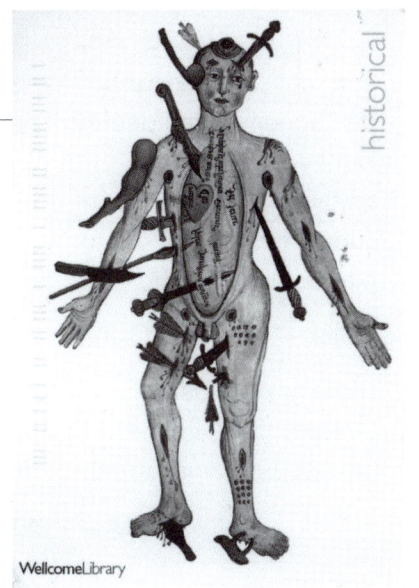

臓器とその痛み Organs and their pains

器官		痛みの名称			和訳
英語名詞	英語形容詞	-algia	-dynia	一般用語	
heart	cardiac	cardi**algia**	cardio**dynia**	heart pain	心臓痛[*1]
nose	nasal, rhinal	rhin**algia**	rhino**dynia**	nose pain, hurt nose, sore nose	鼻痛
mouth	oral	stomat**algia**	stomato**dynia**	-	口腔痛
larynx	laryngeal	laryng**algia**	-	sore throat,[*2] lost voice	喉頭痛
trachea	tracheal	trache**algia**	-	sore throat	気管痛
pleura	pleural	cost**algia**, pleur**algia**	pleuro**dynia**	sore chest, chest cold	肋骨痛、胸膜痛
teeth	dental	dent**algia**, odont**algia**	odonto**dynia**	toothache	歯痛
tongue	glossal, lingual	gloss**algia**	glosso**dynia**	-	舌痛
esophagus	esophageal	esophag**algia**	esophago**dynia**	-	食道痛
stomach	gastric	gastr**algia**	gastro**dynia**	stomachache, belly pain, bellyache, pain in the gut	胃痛
intestine, enteron	intestinal, enteric	enter**algia**	entero**dynia**	-	腸痛
anus, rectum	anal, renal	proct**algia**[*3]	procto**dynia**	-	肛門・直腸痛
liver	hepatic	hepat**algia**	-	-	肝臓痛

臓器名		痛みの名称			
英語名詞	英語形容詞	-algia	-dynia	一般用語	和訳
pancreas	pancreatic	pancreat**algia**	-	-	膵痛
peritoneum	peritoneal	peritone**algia**	-	-	腹膜痛
kidney	renal	nephr**algia**	-	-	腎臓痛
urethra	urethral	urethr**algia**	urethro**dynia**	-	尿道痛
(urinating)		-	uro**dynia**	-	（排尿痛）
bladder	cystic	cyst**algia**	cysto**dynia**	-	膀胱痛
testis	testicular	orchi**algia**, test**algia**	orchio**dynia**	-	精巣痛
penis	penile	phall**algia**	phallo**dynia**	-	陰茎痛
uterus	uterine	hyster**algia**	hystero**dynia**	-	子宮痛
(menstrualtion)	menstrual	menorrh**algia**	-	-	（月経痛）
vagina	vaginal	colp**algia**	colpo**dynia**, vagino**dynia**	-	膣痛
mammary gland	mammal	mast**algia**, mamm**algia**	masto**dynia**	-	乳房痛
bone	osteal, osseous	ost**algia**	osteo**dynia**	sore bones	骨痛
joint	articular	arthr**algia**	arthro**dynia**	sore joints, joint aches, joint pain	関節痛
coxae, hip joint	coxal	cox**algia**	coxo**dynia**	-	股関節痛
muscle	muscular	my**algia**	myo**dynia**	aching muscles, muscle pain, muscle ache(s)*4,	筋肉痛
nerve	nervous	neur**algia**	neuro**dynia**	-	神経痛
sciatic nerve	-	ischi**algia**, ischioneur**algia**	-	-	坐骨神経痛
skin	dermal	dermat**algia**	dermato**dynia**	-	皮膚痛
eye	ocular, ophthalmic	ophthalm**algia**	oculo**dynia**	eye ache	眼(球)痛
ear	otic	ot**algia**	oto**dynia**	ear ache	耳痛
brain	encephalic	encephal**algia**, cephal**algia**	encephalo-**dynia**, cephalo**dynia**	headache *5	頭痛

臓器名でないものは（ ）で示す。

*1 heartache と heartburn：heartache は、心臓痛という症状よりむしろ悲嘆、心の痛みという意味合いになる。これと似ているのが heartburn。胸焼けという症状を表すが、これに ing がついて heartburning となると、不満、不平、ねたみ、嫉妬、恨みという意味に変わる。
*2 sore throat：腫れて声がかすれることを lost voice と表現する。I lost my voice.
*3 proctalgia の proct(o)- は proctodeum（肛門管）より。直腸肛門との関係を表す接頭辞である。
*4 muscle ache：stomachache や headache は、臓器名と ache が合体して一語となっているが、muscle ache は 2 語である。
*5 the brain and head（脳と頭）：日本語でも「脳が痛い」などと言わずに、「頭が痛い」と言うが、英語も同じで一般に口語では、脳と頭は同じ意味で使うことが多い。

Column 痛みを表す用語

■Pain（痛み、疼痛）：痛みを表す一般的名詞である。形容詞は painful。
　　I have been in constant pain.（傷がずっと痛い）
　　My injury is painful.（傷が痛い）
■Sore（痛み、触痛、圧痛）：名詞としての意味もあるが、形容詞として使う方が一般的である。名詞としては、soreness という表現もある。
　　The wounds are sore.（傷が痛い）
■Ache（ずきずき痛む、うずく）：名詞（接尾辞）にも動詞にもなり得る。
　　I have a toothache.（歯痛がある、歯が痛い）というように名詞としても使え、また My tooth aches.（歯が痛む）というように動詞としても使える。

Exercises 3

問1．次の臓器の炎症を表す用語を、接尾辞 -itis を用いて作りなさい。

①胃炎（gastric）＿＿＿＿＿＿＿＿＿＿＿＿＿＿
②気管支炎（bronchial）＿＿＿＿＿＿＿＿＿＿＿＿＿＿
③大腸炎（colic）＿＿＿＿＿＿＿＿＿＿＿＿＿＿
④鼻炎（rhinal）＿＿＿＿＿＿＿＿＿＿＿＿＿＿
⑤脳炎（encephalic）＿＿＿＿＿＿＿＿＿＿＿＿＿＿
⑥膵炎（pancreatic）＿＿＿＿＿＿＿＿＿＿＿＿＿＿
⑦肝炎（hepatic）＿＿＿＿＿＿＿＿＿＿＿＿＿＿

問2．次の用語の左右を正しく結びなさい。

①膿尿　　　　　・　　　　　・ a) pyothorax
②化膿　　　　　・　　　　　・ b) pyuria
③胆嚢蓄膿症　　・　　　　　・ c) pyopoiesis
④膿胸　　　　　・　　　　　・ d) empyema of gallbladder
⑤子宮蓄膿症　　・　　　　　・ e) suppuration
⑥膿形成　　　　・　　　　　・ f) pyometra

問3．下の語群から（　　）に適切なものを選びなさい。

炎症は、英語で（①　　　　）と言う。これはラテン語の「火をつける」という言葉に由来する。炎症は通常、発赤（②　　　　）、熱（③　　　　）、腫脹（④　　　　）、痛み（⑤　　　　）、それに機能の消失（⑥　　　　）を伴う。

a) pain　　b) swelling　　c) loss of function　　d) redness　　e) heat　　f) inflammation

問4．以下の英語の病名と日本語を正しく結びなさい。

① feline infectious peritonitis　　　　　・　　　・a) 犬伝染性肝炎
② swine atrophic rhinitis　　　　　　　・　　　・b) 豚伝染性胃腸炎
③ traumatic pericarditis　　　　　　　・　　　・c) 歯肉炎
④ infectious bovine rhinotracheitis　　・　　　・d) 猫伝染性腹膜炎
⑤ feline viral rhinotracheitis　　　　　・　　　・e) 創傷性心嚢炎
⑥ avian infectious laryngotracheitis　・　　　・f) 豚の萎縮性鼻炎
⑦ traumatic reticulitis　　　　　　　　・　　　・g) 鶏の伝染性喉頭気管炎
⑧ transmissible gastroenteritis　　　　・　　　・h) 牛伝染性鼻気管炎
　　(TGE) in swine　　　　　　　　　　　　　　　・i) 猫ウイルス性鼻気管炎
⑨ gingivitis　　　　　　　　　　　　　・　　　・j) 創傷性第二胃炎
⑩ infectious canine hepatitis　　　　　・

問5．次の臓器の痛みを表す用語を、接尾辞 -algia を用いて作りなさい。

①肝臓痛（hepatic）＿＿＿＿＿＿＿＿＿＿＿＿＿＿
②膵臓痛（pancreatic）＿＿＿＿＿＿＿＿＿＿＿＿＿
③尿道痛（urethral）＿＿＿＿＿＿＿＿＿＿＿＿＿＿
④神経痛（nervous）＿＿＿＿＿＿＿＿＿＿＿＿＿＿

問6．次の臓器の痛みを表す用語を、接尾辞 -dynia を用いて作りなさい。

①心臓痛（cardio-）＿＿＿＿＿＿＿＿＿＿＿＿＿＿
②膀胱痛（cysto-）＿＿＿＿＿＿＿＿＿＿＿＿＿＿＿
③筋肉痛（myo-）＿＿＿＿＿＿＿＿＿＿＿＿＿＿＿
④舌痛（glosso-）＿＿＿＿＿＿＿＿＿＿＿＿＿＿＿

Exercises 3 の答え

問1．① gastritis　② bronchitis　③ colitis　④ rhinitis　⑤ encephalitis　⑥ pancreatitis　⑦ hepatitis
問2．① b)　② e)　③ d)　④ a)　⑤ f)　⑥ c)
問3．① f)　② d)　③ e)　④ b)　⑤ a)　⑥ c)
問4．① d)　② f)　③ e)　④ h)　⑤ i)　⑥ g)　⑦ j)　⑧ b)　⑨ c)　⑩ a)
問5．① hepatalgia　② pancreatalgia　③ urethralgia　④ neuralgia
問6．① cardiodynia　② cystodynia　③ myodynia　④ glossodynia

Chapter 4

細胞の基本構造

The Structure of the Cell

1 細胞と組織 Cells and Tissues

細胞学 Cytology

cell	細胞
cellular	細胞の
intracellular	細胞内の
extracellular	細胞外の
protoplasm	原形質
cytosol	細胞質ゾル、サイトゾル
cytoplasmic organelles	細胞小器官

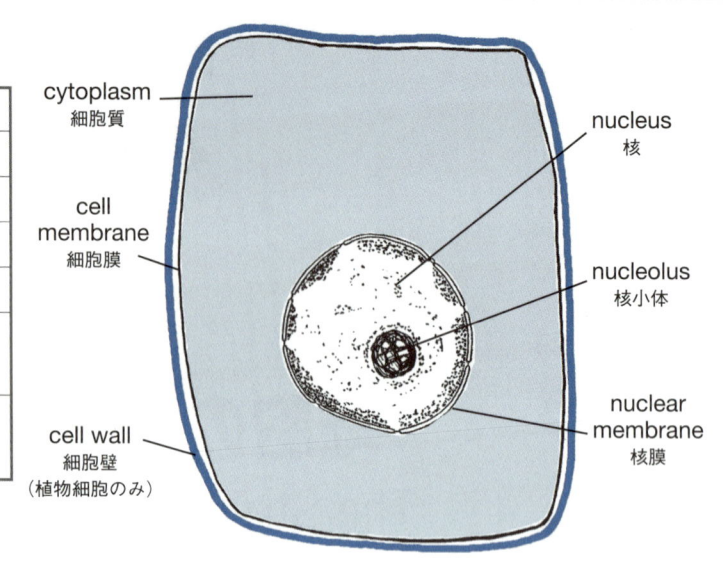

核 Nucleus

語幹	英語名詞	英語形容詞	ラテン語名詞	ギリシャ語名詞
nucl-, nucleo-, karyo-, caryo-	nucleus	nuclear	nucleus	karyon*

＊ 堅い木の実、ナッツの意。

核にかかわる用語 Cellular Nucleus Terminology

nuclear	nuclear membrane, nuclear envelope (=**karyo**theca)	核膜
nucleo-	**nucleo**plasm (=nuclear matrix, **karyo**plast)	核質
caryo-	eu**caryo**te (=eu**karyo**te)	真核細胞、真核生物
karyo-	**karyo**lysis	核融解

細胞接着装置 Cell Junctions

adherens junction, macula adherens, desmosome	接着斑（デスモゾーム）
tight junction, zonula occludens	密着帯、閉鎖帯
gap junction, nexus	細隙結合（ネクサス）

細胞質内封入体 Cytoplasmic Inclusions

glycogen granule	グリコーゲン顆粒
lipid droplet, adiposome	脂肪滴（脂肪小体）
pigment granule	色素顆粒

細胞小器官 Cytoplasmic Organelles

🔊

- lysosome ライソゾーム、水解小体
- Gorgi apparatus ゴルジ装置
- cilia 線毛
- microvilli 微絨毛
- secretory vesicle 分泌顆粒
- microfilament マイクロフィラメント、微細糸
- microtubule 微細管
- cytoskeleton 細胞骨格
- rough endoplasmic reticulum (rER) 粗面小胞体
- ribosome リボゾーム
- smooth endoplasmic reticulum (sER) 滑面小胞体
- centriole 中心子
- mitochondria ミトコンドリア、糸粒体

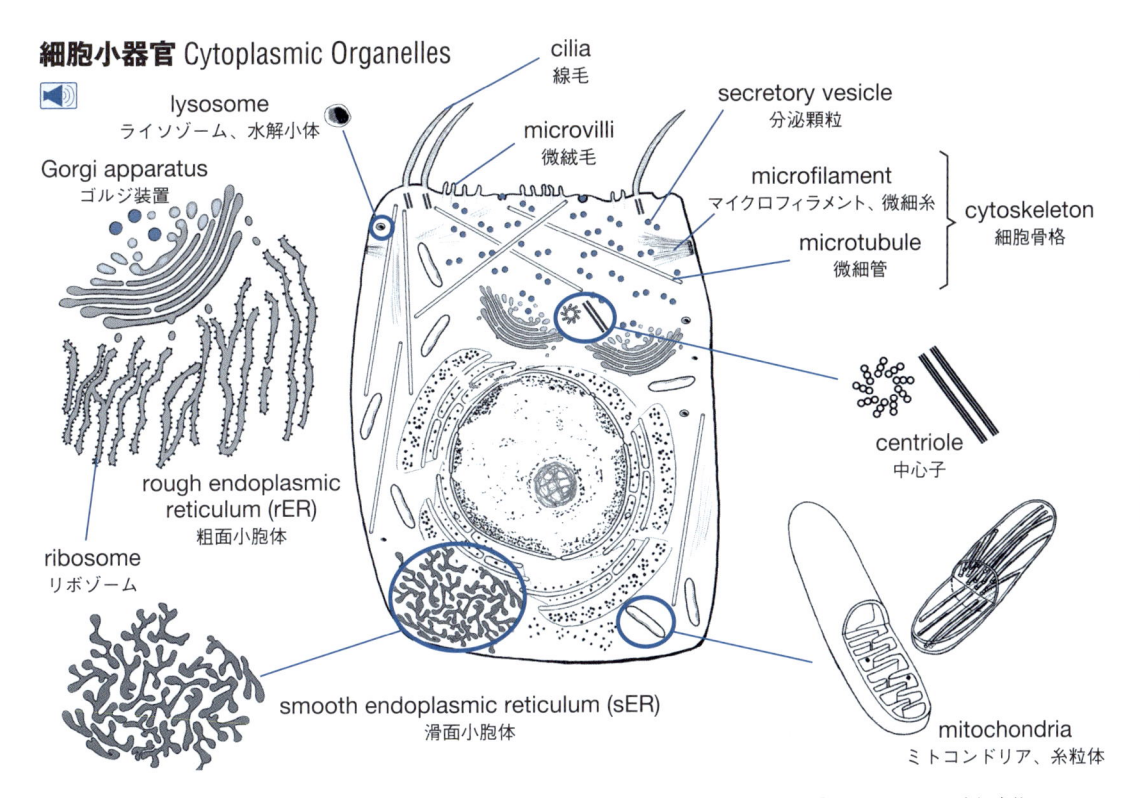

ライソゾーム（リソゾーム）lysosome：lyso はギリシャ語の lysis（溶解）+ some（soma［体］）より。和訳は水解小体。

ゴルジ装置 Golgi apparatus、Golgi complex：解剖学者 Golgi の名から命名された。以前は多くの解剖学用語が発見者の名前で呼ばれていたが、次第に一般名称に置き換えられつつある。そのような流れのなかで、固有名詞を保っているほとんど唯一の解剖学用語でもある。

小胞体 reticulum：ラテン語の rete（網）+ culum（指小辞）。「小さい網」という意味。

線毛 cilium (sg.)、cilia (pl.)：cilia はラテン語でまぶたのこと。しかし17世紀になって解剖学者がこの言葉をまぶたではなく、まつ毛の意味で使い始めた。その後、顕微鏡の発明により線毛という意味で使われるようになった。

微絨毛 microvillus (sg.)、microvilli (pl.)：villus はラテン語で、動物のもしゃもしゃした毛のこと。なお、cilia より microvilli の方がずっと細く、通常、小さい。

ミトコンドリア mitochondrion (sg.)、mitochondria (pl.)：ギリシャ語の mitos（糸）+ khondrion（顆粒、軟骨）より、和訳は糸粒体。大部分の細胞は層板状のクリスタ（cristae）（図左）を持つが、ステロイド産生細胞は小管状の cristae（図右）を持つ。

Column ナッツ

核のラテン語 nucleus は nux（くるみなどのナッツ）に由来する。nux（nucs）に「小さい」という意味の "le" を挿入して "nucleus（小さな胡桃）" となった。

ラテン語で何か小さいものを表すとき、アルファベットの "l（エル）" を挿入して言葉を作る方法がある。これが英語に反映された例には、artery（動脈）と arteriole（小動脈）、vein（静脈）と venule（小静脈）、cerebrum（大脳）と cerebellum（小脳）などがある。そのひとつが nux と nucleus である。そう、細胞はその芯に小さなナッツを持っているのだ。

続けて、核小体（nucleolus）を見てみよう。これは核（nucleus）に指小辞の "l" がはさまっている。nucleus の更に小さいもの、それが nucleolus なのである。直訳すると、「小さな小さな、とても小さなナッツ」とでも言おうか。

細胞の増殖と死 Cellular Proliferation and Death

proliferation	増殖
multiplication	増殖、増加
mitosis	有糸分裂
meiosis	減数分裂
apoptosis, programmed cell death	アポトーシス（プログラム細胞死）
necrosis	ネクローシス
differentiation	分化
maturation 成熟	mature	成熟した
	immature	未熟な
degeneration	変性
regeneration	再生

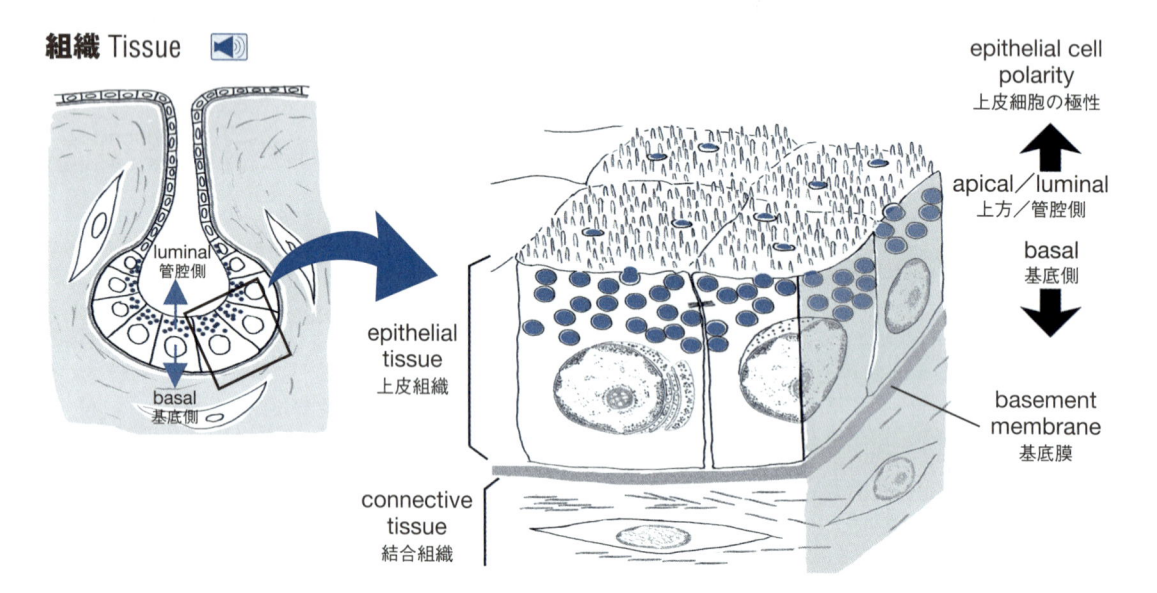

組織 Tissue

（図：epithelial cell polarity 上皮細胞の極性／apical/luminal 上方／管腔側／basal 基底側／basement membrane 基底膜／luminal 管腔側／basal 基底側／epithelial tissue 上皮組織／connective tissue 結合組織）

諸組織 Various Tissues

epithelial tissue 上皮組織	epithelium (sg.), epithelia (pl.)	上皮
	mucosa	粘膜
	gland	腺
connective tissue 結合組織	collagen fiber	膠原線維
	elastic fiber	弾性線維
	reticular fiber	細網線維
basement membrane, basal lamina		基底膜（基底板）
cartilaginous tissue		軟骨組織
osseous tissue		骨組織
muscular tissue		筋組織
nervous tissue		神経組織

2 遺伝学 Genetics

🔊

gene	遺伝子[1]
genetics	遺伝学
geneticist	遺伝学者
gene therapy	遺伝子治療[2]
genetically modified organism (GMO)	遺伝子組み換え生物、遺伝子改変生物[3]
transgenic	遺伝子導入の[4]
genome	ゲノム[5]

[1] 生物の遺伝形質を既定する因子。
[2] 遺伝子を導入し、細胞のDNAを改変する治療。
[3] 動植物を問わない。大豆などのGMOが大きな論争になっている。
[4] トランスジェニックマウスなど。ノックアウト（特定の遺伝子を発現できなくした）、ノックイン（外来遺伝子を人為的に導入した）などがある。
[5] 全遺伝情報。gene + -ome（ラテン語で集合体を意味する接尾辞）より。chromosome（染色体）のomeも同様である。

🔊

deoxyribonucleic acid (DNA)		デオキシリボ核酸
ribonucleic acid (RNA) リボ核酸	messenger RNA (mRNA)	メッセンジャー（伝令）RNA
	ribosomal RNA (rRNA)	リボゾームRNA
	transfer RNA (tRNA)	トランスファー（転位、運搬）RNA
codon		コドン[1]
peptide		ペプチド[2]

[1] 3塩基1組でアミノ酸1個を指定する塩基配列。
[2] アミノ酸数個〜数十個がペプチド結合した化合物。

🔊

replication	複製
transcription	転写[1]
translation	翻訳[2]

[1] DNAの遺伝子情報をmRNAに写すこと。
[2] mRNAの情報をもとに、蛋白質を合成すること。

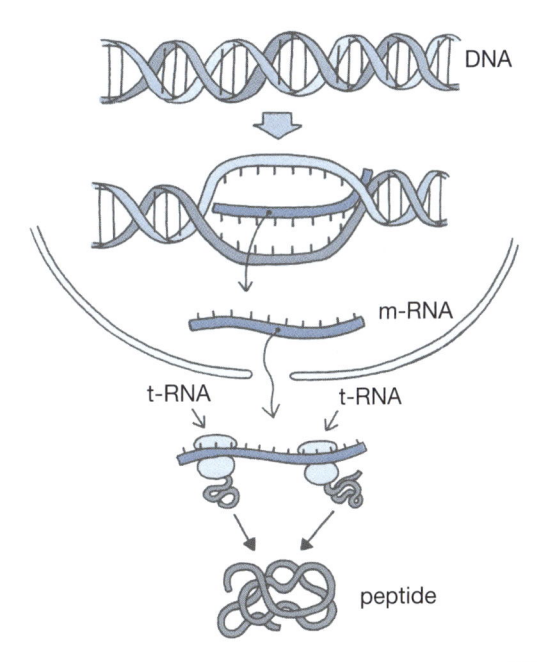

遺伝子解析

PCR：polymerase chain reaction（ポリメラーゼ連鎖反応）。DNAの人工的な増幅法。
RT-PCR：reverse transcription polymerase chain reaction（逆転写ポリメラーゼ連鎖反応）。mRNAを鋳型にcDNAを作り、PCRする方法。
real time-PCR：real time polymerase chain reaction（リアルタイムポリメラーゼ連鎖反応）。PCRでの増幅産物の生成過程を、同時に経時的にモニタリングし、解析する方法。頭文字は上記のRT-PCRと同じだが、内容は全く異なる。
Southern blotting（サザンブロッティング）：DNAのブロッティング法。
northern blotting（ノーザンブロッティング）：RNAのブロッティング法。
western blotting（ウエスタンブロッティング）：蛋白質のブロッティング法。

・ Southernは人名なので、必ず大文字で、northernとwesternは小文字で始める。

　ブロッティングの名称は、Southern、northern、westernとあるが、東を意味するeastern blottingは無い。歴史的にはSouthernが最初で、方角ではなく人名由来である。北と西はこれにちなんだ「しゃれ」である。
　いつかFar eastの国から、easternを提唱する説が出てくるだろうか？

Column アポトーシス

　apoptosisはギリシャ語のapo-「離れる」と、同じくギリシャ語のptosis「落ちる」より合成された造語である。

　秋になると落葉樹の木の葉は枝からひらひら「離れて」（apo）、「落ちる」（ptosis）。葉の基部と枝の先端を結ぶところの細胞が死ぬためである。この死は遺伝的にプログラムされた死で、異常なものではない。また葉が死ぬからといって、木全体が死ぬわけでもない。むしろ晩秋に葉を落とすことにより、冬の木を守ることになる。このように、総体としての個体（動物であれ、植物であれ）を生かすために、その中の一部の細胞が死ぬように遺伝的にプログラムされた死、それがapoptosisである。

　apoptosisという考え方は20世紀末に提唱されたもので、新しい説と言える。しかしこの言葉を提唱した科学者はギリシャ語にさかのぼって、apo + ptosisと命名したわけである。日本語でも新しい言葉を作る時、適切な漢字を選んで、合成することがある。例えば「電話」という言葉は、昔は無かったが、その意味からか適切な漢字が選ばれた。英語でも、しばしばギリシャ語やラテン語にさかのぼって、適切な言葉を探し出し、合成して造語する、ということが行われる。ギリシャ語やラテン語は接頭辞、接尾辞なども豊富で造語しやすく、かつ、ちょっと耳にカッコいい響きがするためだろう。

Chapter 5

筋骨格系とその疾患
The Musculoskeletal System and its Disorders

1 骨格系 The Skeletal System

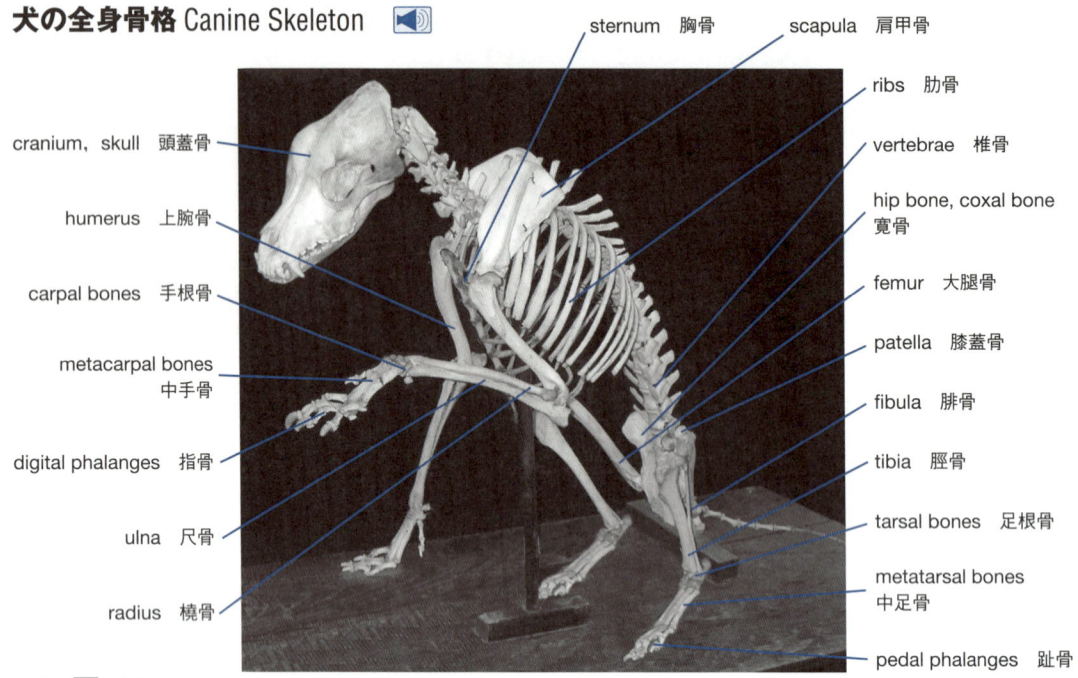

犬の全身骨格 Canine Skeleton

- sternum　胸骨
- scapula　肩甲骨
- ribs　肋骨
- vertebrae　椎骨
- hip bone, coxal bone　寛骨
- femur　大腿骨
- patella　膝蓋骨
- fibula　腓骨
- tibia　脛骨
- tarsal bones　足根骨
- metatarsal bones　中足骨
- pedal phalanges　趾骨
- cranium, skull　頭蓋骨
- humerus　上腕骨
- carpal bones　手根骨
- metacarpal bones　中手骨
- digital phalanges　指骨
- ulna　尺骨
- radius　橈骨

a) 骨 Bone

語幹	英語名詞	英語形容詞	ラテン語名詞	ギリシャ語名詞
oste(o)-, ossi-	bone	osteal	os(sg.), ossa(pl.)	-

骨にかかわる用語 Skeletal Terminology

oste(o)-, ossi-	**oste**on	オステオン（骨単位）
	osteocyte	骨細胞
	osteoblast	骨芽細胞
	osteoclast	破骨細胞
	ossify	骨化する、骨形成する
	peri**osteu**m*	骨膜
	ostealgia, **osteo**dynia	骨痛
	osteitis	骨炎
	pan**oste**itis	汎骨炎
	osteoma	骨腫
	osteotomy	骨切り術
	osteolysis	骨溶解
	osteomyelitis	骨髄炎
	osteoporosis	骨粗鬆症
	osteomalacia	骨軟化症
	osteochondrosis dissecans	離断性骨軟骨症
	osteodystrophy	骨異栄養症

bone		osteopsathyrosis	骨脆弱症
		osteoarthritis	骨関節炎
	bone marrow		骨髄
	bone density		骨密度
	bone trabecula		骨梁
	bone fracture		骨折
	bone ache		骨痛
その他	periosteal reaction		骨膜反応
	rachitis, rickets		くる病

* peri-：「周囲」を表す接頭辞。

b) 椎骨 Vertebrae

vertebral column 脊柱	cervical vertebrae 頚椎	atlas	環椎
		axis	軸椎
	thoracic vertebrae		胸椎
	lumbar vertebrae		腰椎
	sacral vertebrae／sacrum		仙椎／仙骨
	caudal（coccygeal）vertebrae		尾椎

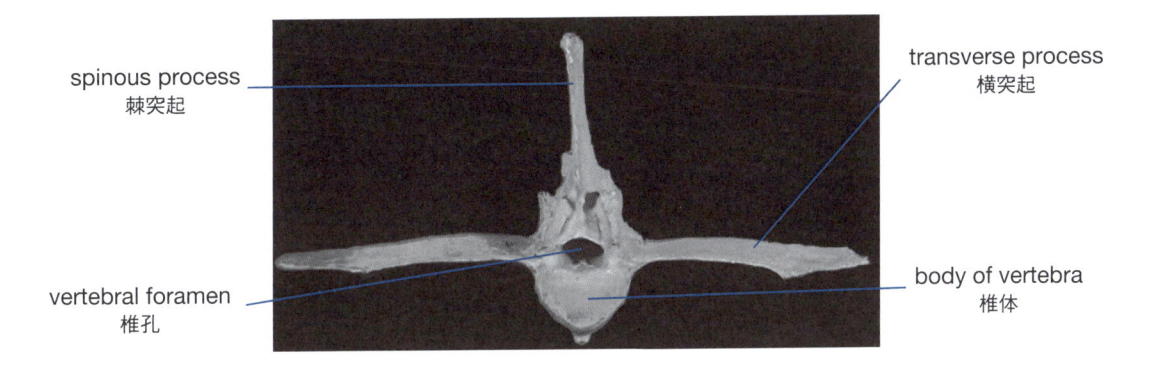

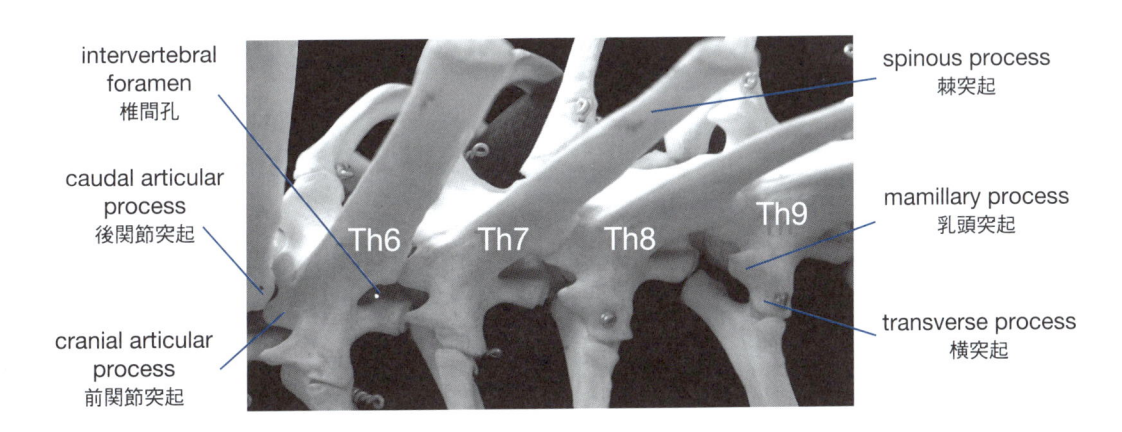

- spinous process 棘突起
- vertebral foramen 椎孔
- transverse process 横突起
- body of vertebra 椎体
- intervertebral foramen 椎間孔
- caudal articular process 後関節突起
- cranial articular process 前関節突起
- spinous process 棘突起
- mamillary process 乳頭突起
- transverse process 横突起

Th6　Th7　Th8　Th9

c) 前肢骨格 Skeleton of the Forelimb (Thoracic Limb) 🔊

scapula		肩甲骨
clavicle		鎖骨
skeleton of upper arm 上腕骨格	humerus	上腕骨
skeleton of forearm (antebrachium) 前腕骨格	radius	橈骨
	ulna	尺骨
skeleton of manus 手骨格	carpal bones	手根骨
	metacarpal bones	中手骨
	digital phalanges	指骨

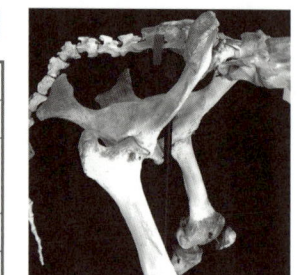

牛の骨盤（斜め側方より）

d) 後肢骨格 Skeleton of the Hindlimb (Pelvic Limb) *1 🔊

hip bone, coxal bone 寛骨	ilium	腸骨
	ischium	坐骨
	pubis	恥骨
skeleton of thigh 大腿骨格	femur	大腿骨
	patella	膝蓋骨
skeleton of leg (cruris) 下腿骨格	tibia	脛骨
	fibula *2	腓骨
skeleton of pes (hindpaw) 足骨格	tarsal bones	足根骨
	metatarsal bones	中足骨
	pedal phalanges	趾骨

馬の骨盤（頚側より）

＊1　骨盤は pelvis、寛骨臼は acetabulum。
＊2　形容詞は fibular または peroneal。

e) 指骨 Digital Phalanges 🔊

すべての動物に共通		有蹄類（ungulata）のみ		食肉類（carnivore）のみ	
proximal phalanx	基節骨	pastern bone	繋骨	—	—
middle phalanx	中節骨	small pastern bone	冠骨	—	—
distal phalanx, terminal bone	末節骨	coffin bone	蹄骨	unguiculate bone	鈎爪骨

馬の中手骨、指骨および蹄（蹄は骨に分類されない）

Column アトラス

　第一頚椎（環椎）の愛称はアトラス（Atlas）という。アトラスはギリシャ神話の巨人で、ゼウスの命により重い天空を支えている。この神話より、重い頭蓋骨を支える第一頚椎のことをそう呼ぶようになった。また、昔の地図帳には天球を両肩の間と後頚部で支えるアトラスの絵が掲載されていたことから、地図帳や図版集をも Atlas と呼ぶならわしとなった。

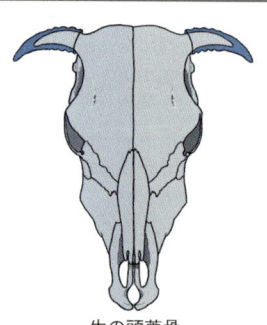

牛の頭蓋骨
（角は皮膚が特殊化したものであり、骨に分類されない）

よくある骨の疾患 Diseases of the Bone

bone fracture	骨折
dislocation	脱臼
subluxation	亜脱臼
coxofemoral luxation	股関節脱臼
patellar luxation	膝蓋骨脱臼
hip dysplasia	股関節形成不全

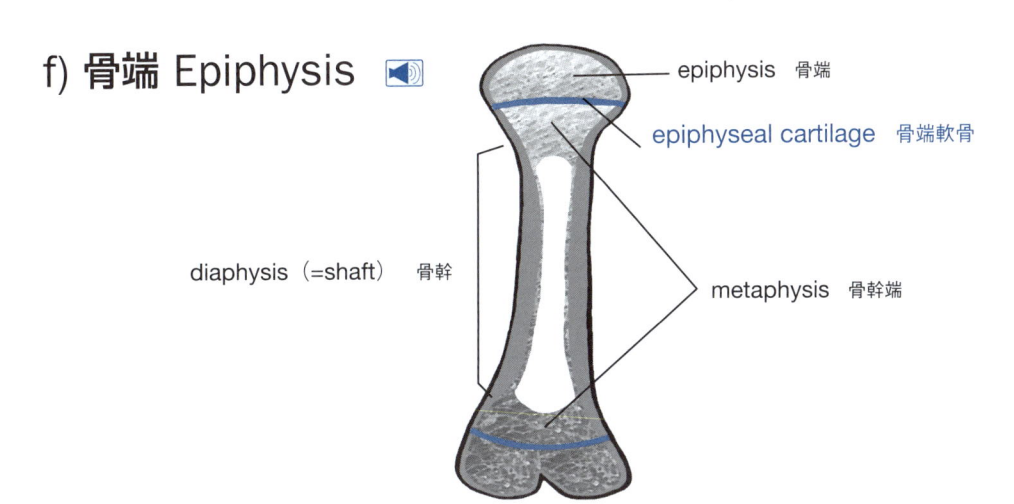

f) 骨端 Epiphysis

- epiphysis 骨端
- **epiphyseal cartilage** 骨端軟骨
- diaphysis (=shaft) 骨幹
- metaphysis 骨幹端

-physis はギリシャ語の「成長」を意味し、骨の成長にかかわる用語として使われる。symphysis pubica（恥骨結合、骨盤結合）の symphysis も sym-（共に）＋ -physis（成長）に由来する。ちなみに hypophysis（下垂体）も hypo-（下へ）＋ -physis（成長）に由来する。

g) 骨髄 Bone Marrow

語幹	英語名詞	英語形容詞	ラテン語名詞	ギリシャ語名詞
myel(o)-	bone marrow	myelic	medulla ossium	myelos

myel- には、marrow（髄）という意味しかない。すなわち、骨髄も脊髄も、語幹は myel- であり、myelencephalon（髄脳〔延髄〕）も myel- を含む。従って髄炎（myelitis）は、骨髄炎と脊髄炎の両方を指すので、注意が必要である。

骨髄にかかわる用語 Bone Marrow Terminology

myel(o)-	**myel**itis	骨髄炎（または脊髄炎）
	myeloma	骨髄腫
	osteo**myel**itis	骨髄炎
	multiple **myelo**ma	多発性骨髄腫
	myeloblast	骨髄芽球
	myeloblastoma	骨髄芽球腫
	myelosis	骨髄症

h) 軟骨 Cartilage

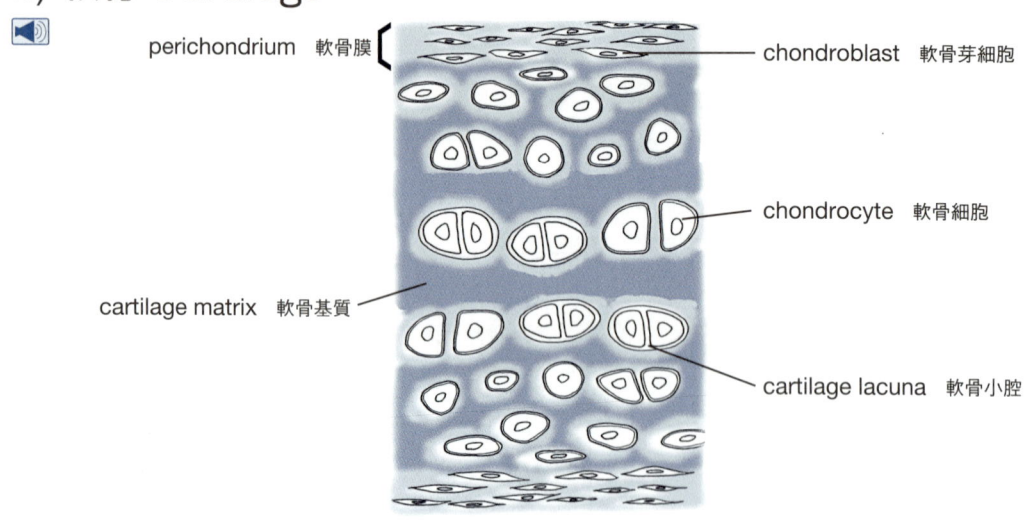

图中标注:
- perichondrium 軟骨膜
- chondroblast 軟骨芽細胞
- chondrocyte 軟骨細胞
- cartilage matrix 軟骨基質
- cartilage lacuna 軟骨小腔

語幹	英語名詞	英語形容詞	ラテン語名詞	ギリシャ語名詞
chondr(o)-	cartilage	chondral	cartilago	chondros

軟骨にかかわる用語 Cartilage Tissue Terminology

chondr(o)-	**chondro**cyte	軟骨細胞
	peri**chondr**ium	軟骨膜
	chondralgia, **chondro**dynia	軟骨痛
	chondritis	軟骨炎
	peri**chondr**itis	軟骨膜炎
	chondroid	軟骨様の
cartilage	cartilaginoid	軟骨様の
	hyaline cartilage[*1]	硝子軟骨
	fibrocartilage	線維軟骨
	elastic cartilage[*2]	弾性軟骨

[*1] hyalin（硝子質）はギリシャ語の hyalos（硝子）に由来する。
[*2] elastic fibers（弾性線維）を多量に含む。

Column のど仏

　生きている人の「のど仏」と呼ばれるのは甲状軟骨である。しかし亡くなって火葬すると、軟骨は焼けてしまって残らない。甲状軟骨のあたりに残るのは第二頸椎（axis）である。axis の歯突起がちょうど仏様の頭、椎体が仏様の体、椎弓が座っている仏様の手足のように見えるため、火葬場の人は第二頸椎を指して、「これがのど仏です」と説明する。

2 関節 The Joint

サルの主な関節 Main Joints of the Monkey

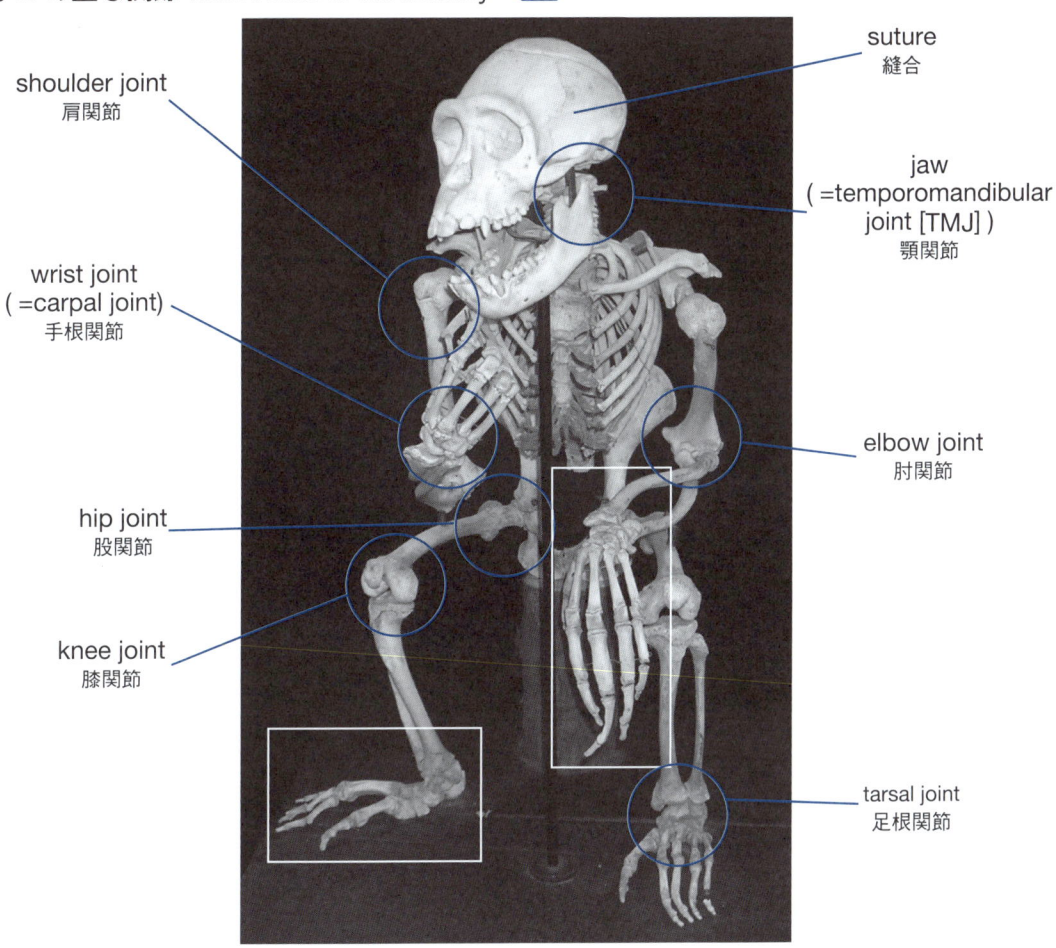

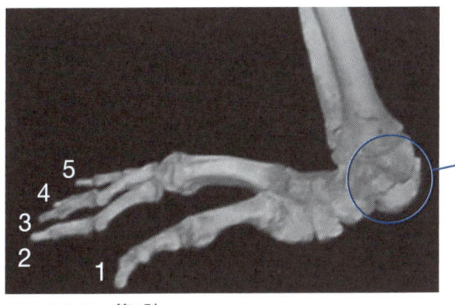

- shoulder joint 肩関節
- wrist joint (=carpal joint) 手根関節
- hip joint 股関節
- knee joint 膝関節
- suture 縫合
- jaw (=temporomandibular joint [TMJ]) 顎関節
- elbow joint 肘関節
- tarsal joint 足根関節

heel かかと

1：digit I　第1趾
2〜4：digit II - digit IV　第2〜4趾
5：digit V　第5趾

1：digit I　第1指
2：digit II　第2指
3：digit III　第3指
4：digit IV　第4指
5：digit V　第5指

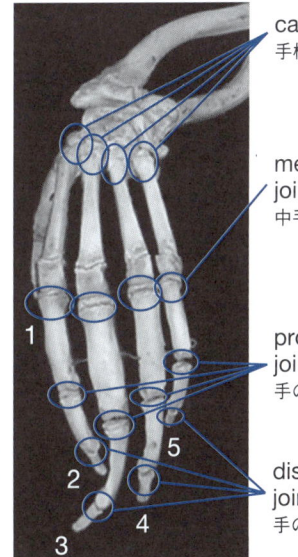

- carpometacarpal joints 手根中手関節
- metacarpophalangeal joints 中手指節関節
- proximal interphalangeal joints of manus 手の近位指節間関節
- distal interphalangeal joints of manus 手の遠位指節間関節

関節 Joint

語幹	英語名詞	英語形容詞	ラテン語名詞	ギリシャ語名詞
arthr(o)-	joint, articulation	articular	articulatio	arthron

　関節は、synovial joint（滑膜性関節）：肩関節、股間節、膝関節、fibrous joint（線維性関節）：頭蓋、歯槽、cartilaginous joint（軟骨性関節）：（左右の）下顎骨の結合、成長板などに分類される。

日本カモシカの全身骨格と主な関節
Skeleton and Main Joints of the Japanese Serow

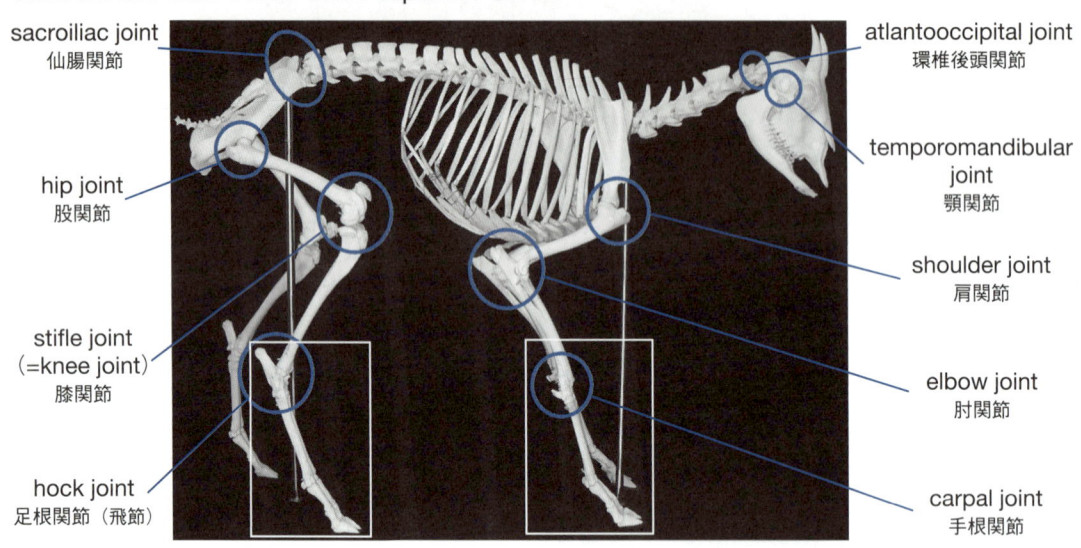

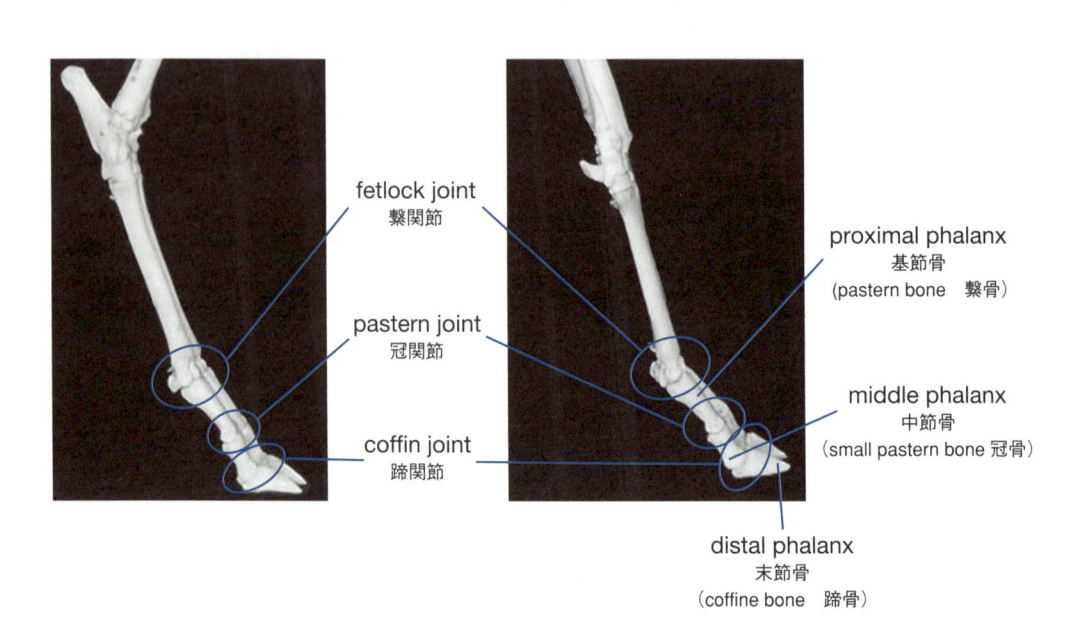

- sacroiliac joint 仙腸関節
- hip joint 股関節
- stifle joint (=knee joint) 膝関節
- hock joint 足根関節（飛節）
- atlantooccipital joint 環椎後頭関節
- temporomandibular joint 顎関節
- shoulder joint 肩関節
- elbow joint 肘関節
- carpal joint 手根関節
- fetlock joint 繋関節
- pastern joint 冠関節
- coffin joint 蹄関節
- proximal phalanx 基節骨 (pastern bone 繋骨)
- middle phalanx 中節骨 (small pastern bone 冠骨)
- distal phalanx 末節骨 （coffine bone 蹄骨）

関節にかかわる用語 Joint Terminology

arthr(o)-	**arthro**plasty	関節形成術
	arthrodesis	関節固定術
	arthritis	関節炎
	arthrosteitis	骨関節炎
	arthrosis	関節症 *1
	arthrotomy	関節切開術
	arthropyema	関節腫
その他	articular fracture	関節骨折
	ankylosis	関節の強直症 *2

*1 関節の非炎症性変化を指す。
*2 線維性あるいは骨性に結合し硬直する疾病。

滑膜性関節の構造とその疾患
Structure of the Synovial Joint and their Disorders

articular capsule, joint capsule	関節包
synovia, synovial fluid	滑液
synovial bursa	滑液包
synovial membrane	滑膜
synovial sarcoma	滑膜肉腫
synovioma*	滑膜腫
bicipital tenosynovitis*	二頭筋腱滑膜炎
cranial cruciate ligament rupture	前十字靭帯断裂

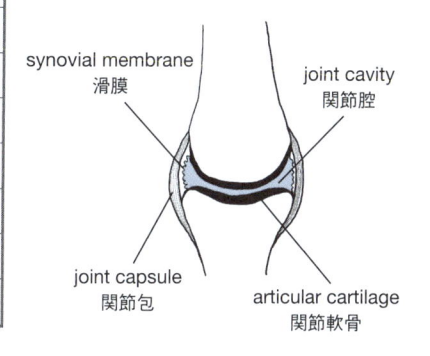

synovial membrane 滑膜
joint cavity 関節腔
joint capsule 関節包
articular cartilage 関節軟骨

* synovi- は synobial membrane（滑膜）の語幹である。

関節の疾患 Joint Disorders

joint mouse (sg.), joint mice (pl.)		関節鼠（関節遊離体）*
intervertebral disk 椎間板（椎間円板）	discitis	椎間板炎
	intervertebral disk protrusion （=hernia of intervertebral disk）	椎間板ヘルニア
spondylosis deformans		変形性脊椎症
ankylosing spondylosis（AS）		強直性脊椎炎
hip dysplasia		股関節形成不全、股異形成
nonunion（pseudoarthrosis）		癒合不全（偽関節）
osteochondritis dissecans（OCD lesion）		離断性骨軟骨炎
sprain		捻挫
whiplash injury		むち打ち症

* 関節内に剥がれ落ちた骨や軟骨の一部がネズミのようにチョロチョロ動くことに由来する。

3 筋 The Muscle System

筋 Muscle

語幹	英語名詞	英語形容詞	ラテン語名詞	ギリシャ語名詞
myo-, mys(i)-, sarc(o)-	muscle, flesh	muscular	musculus	mys, myos sarx

　ギリシャ語の mys、ラテン語の musculus は、解剖学用語としての筋肉という意味である。筋肉物質、肉類似物を表すギリシャ語 sarx に由来する sarco- は、筋を意味する接頭辞であると同時に、上皮ではなく間葉系由来という意味を持つ接頭辞となった。sarx と似た意味を持つ英語は flesh であり、ゲルマン語の flesch に由来する。flesh は肉を含む、軟部組織（骨以外）という意味。なお、meat は、食用としての肉を意味する。

筋にかかわる用語 Muscle Tissue Terminology

語幹	用語	和訳
myo-	**myo**logy	筋学
	myocyte	筋細胞
	myoblast	筋芽細胞
	myofibril	筋原線維
	myoglobulin	ミオグロブリン
	myositis	筋炎
	myoma	筋腫
	myopathy	筋症、筋障害
	myoatrophy	筋萎縮
	myocarditis	心筋炎
mys(i)-	endo**mysi**um	筋内膜
	peri**mysi**um	筋周膜
	peri**mysi**itis, erimysitis（=peri**myo**sitis）	筋周膜炎
	epi**mysi**um	筋上膜
sarc(o)-	**sarco**lemma	筋細胞膜
	sarcomere	筋節
	sarcoplasm	筋形質
	sarcoplasmic reticulum	筋小胞体

> **Column　マウスと筋肉**
>
> 　筋肉を意味するラテン語の"musculus"はマウスに由来する。マウスの学名（ラテン語）は *Mus musculus* である。なぜ筋肉がマウスなのだろうか？一説には、上腕二頭筋（力こぶを作るとき、盛り上がる筋肉）の形がマウスに似ているからというもの、また一説には皮膚の下で収縮したり弛緩したりする筋肉の動きが、敷物の下をチョロチョロと動くマウスの動きと似ているからだという。

筋、筋膜、腱 Muscles, Fasciae and Tendons 🔊

smooth muscle		平滑筋
striated muscle	cardiac muscle	心筋
横紋筋	skeletal muscle	骨格筋
involuntary muscle		不随意筋*1
voluntary muscle		随意筋*2
fascia (sg.)*3, fasciae, fascias (pl.)		筋膜
tendon	synovial sheath of tendon	腱の滑液鞘
腱	tendinous sheath	腱鞘
aponeurosis*4		腱膜
aponeurositis		腱膜炎
ligament*5		靭帯

*1　不随意筋には平滑筋、心筋が含まれる。
*2　随意筋には骨格筋が含まれる。
*3　fasciaは同じつづりのラテン語 fascia（平らで細長いもの）に由来。
*4　ギリシャ語の apo-（離れる）+ neuron（腱）より。neuron というと神経細胞を連想するが、古代ギリシャ語の neuron には、「腱、糸、神経などの細くて長いもの」の意味があり、これが次第に神経に限局して用いられるように変化した。この aponeurosis という言葉は、neuron を腱という意味で使っており、「腱からしだいに離れた（離れて膜状になる）」の意味。
*5　ラテン語の ligamentum（結合するもの）より。英語の ligate（結紮する）、ligature（ひも、縛る）と同じ語源である。

筋の部位 Parts of the Muscle 🔊

head of muscle	筋頭
belly of muscle（=muscle belly）	筋腹
tail of muscle（=muscle tail）	筋尾
origin	起始
insertion	終止
cutaneous muscle	皮筋

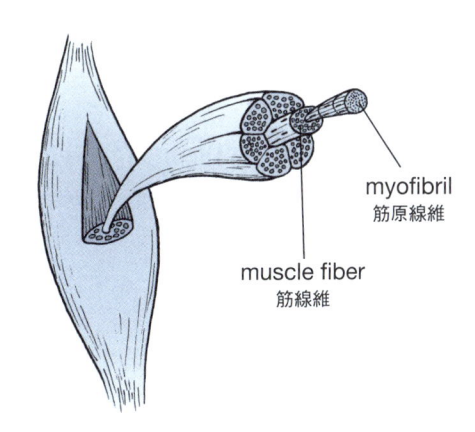

myofibril 筋原線維
muscle fiber 筋線維

> **Column　アキレス腱**
>
> ギリシャ神話の英雄アキレス（Achilleus）の母は、まだ赤ん坊のアキレスを冥府の川（Styx）の水に浸した。川の中に赤ん坊が落ちてしまわぬよう、母親はアキレス腱のところを持って、逆さに水の中に浸した。Styx の水に浸かった体はほとんど不死となり、アキレスは長じてトロイ戦争で大活躍する英雄となる。けれども最後に唯一、Styx の水に浸っていなかった泣き所、アキレス腱（Achilles tendon）を矢で射られて、倒され、殺されてしまうのだ。

筋の機能 Muscular Function 🔊

機能		対応する筋の種類	
flexion	屈曲	flexor muscle	屈筋
extension	伸展	extensor muscle	伸筋
abduction	外転	abductor muscle	外転筋
adduction	内転	adductor muscle	内転筋
outward rotation／inward rotation	外旋／内旋	rotator muscle	回旋筋
pronation	回内	pronator muscle	回内筋
supination	回外	supinator muscle	回外筋
dilation	散大	dilator muscle	散大筋
drawing close	括約	sphincter muscle	括約筋
lifting up	挙上	levator muscle	挙筋
depression	下制	depressor muscle	下制筋

緊張 Tension 🔊

tension	緊張、伸長
contraction	収縮
atony	無緊張、緊張減退、アトニー

筋肉の疾患 Muscle Disorders 🔊

convulsion		ひきつけ（全身の痙攣）
twitch, cramp		（足の筋肉などの）痙攣
tic		（顔面などの）痙攣
strain, pulled muscle		肉離れ*、筋挫傷
muscle contusion		筋挫傷
laceration		（筋－腱単位内の）裂傷
muscle contracture		筋肉の拘縮
myositis 筋炎	muscular rheumatism	筋肉リウマチ
	polymyositis	多発性筋炎
	dermatomyositis	皮膚筋炎
	suppurative myositis	化膿性筋炎
myopathy ミオパチー、筋病	muscular dystrophy	筋ジストロフィー
	rhabdomyolysis	横紋筋融解［症］
	myasthenia gravis	重症筋無力症
degenerative muscular disease		筋肉の退行性の疾病

* 「彼は足に肉離れをおこした」は "He tore a muscle in his leg." "His muscle was torn." のように表される。

問1．左右を正しく結びなさい。

1. ①骨細胞　　　　・　　　　・a) osteoclast
   ②骨芽細胞　　　・　　　　・b) osteocyte
   ③破骨細胞　　　・　　　　・c) chondroblast
   ④軟骨細胞　　　・　　　　・d) osteoblast
   ⑤軟骨芽細胞　　・　　　　・e) chondrocyte

2. ①滑膜　　　　　・　　　　・a) synovia
   ②滑液　　　　　・　　　　・b) articular capsule
   ③関節包　　　　・　　　　・c) synovial membrane

3. ①筋　　　　　　・　　　　・a) aponeurosis
   ②筋膜　　　　　・　　　　・b) fascia
   ③腱　　　　　　・　　　　・c) ligament
   ④腱膜　　　　　・　　　　・d) muscle
   ⑤靭帯　　　　　・　　　　・e) tendon

4. ①骨折　　　　　・　　　　・a) dislocation
   ②疲労骨折　　　・　　　　・b) intervertebral disc hernia
   ③偽関節　　　　・　　　　・c) sprain
   ④脱臼　　　　　・　　　　・d) fatigue fracture
   ⑤捻挫　　　　　・　　　　・e) dysosteogenesis
   ⑥骨形成不全　　・　　　　・f) fracture
   ⑦椎間板ヘルニア・　　　　・g) pseudoarthrosis, nonunion, falty union

5. ①骨髄炎　　　　・　　　　・a) arthrosteitis
   ②骨腫瘍　　　　・　　　　・b) frozen shoulder
   ③骨肉腫　　　　・　　　　・c) osteosarcoma
   ④軟骨肉腫　　　・　　　　・d) chondrosarcoma
   ⑤関節リウマチ　・　　　　・e) myelitis
   ⑥骨軟化症　　　・　　　　・f) osteomalacia
   ⑦骨粗鬆症　　　・　　　　・g) rheumatoid arthritis
   ⑧五十肩　　　　・　　　　・h) osteoporosis
   ⑨骨関節炎　　　・　　　　・i) bone tumor

問２．次の用語を英語になおしなさい。

1. oste(o)- とつないで次の用語を作りなさい。
   ①骨炎＿＿＿＿＿＿＿＿＿＿＿＿＿＿
   ②骨症＿＿＿＿＿＿＿＿＿＿＿＿＿＿
   ③骨切除術＿＿＿＿＿＿＿＿＿＿＿＿＿
   ④骨形成術＿＿＿＿＿＿＿＿＿＿＿＿＿
   ⑤骨芽細胞＿＿＿＿＿＿＿＿＿＿＿＿＿

2. arthr(o)- とつないで次の用語を作りなさい。
   ①関節痛＿＿＿＿＿＿＿＿＿＿＿＿＿＿
   ②関節炎＿＿＿＿＿＿＿＿＿＿＿＿＿＿
   ③関節切除術＿＿＿＿＿＿＿＿＿＿＿＿
   ④関節形成術＿＿＿＿＿＿＿＿＿＿＿＿
   ⑤関節解離＿＿＿＿＿＿＿＿＿＿＿＿＿

3. myo- とつないで次の用語を作りなさい。
   ①筋炎＿＿＿＿＿＿＿＿＿＿＿＿＿＿
   ②筋学＿＿＿＿＿＿＿＿＿＿＿＿＿＿
   ③筋芽細胞＿＿＿＿＿＿＿＿＿＿＿＿＿
   ④筋電図＿＿＿＿＿＿＿＿＿＿＿＿＿＿
   ⑤筋壊死＿＿＿＿＿＿＿＿＿＿＿＿＿＿

Column　肉の部位

牛肉の名称を英語で紹介しよう。
- カルビ：肋間筋（intercostal muscle）
- サーロイン：腰最長筋（lumbar longissimus muscle）
- ヒレ：大腰筋（psoas major muscle）
- 三枚肉：腹筋（abdominal muscles）＝外腹斜筋（external oblique abdominal muscle）、内腹斜筋（internal oblique abdominal muscle）、腹横筋（transverse abdominal muscle）
- ハラミ、サガリ：横隔膜（diaphragm）の筋部。
- タン：舌（tongue）
- ハツ：心臓（heart）
- マメ：腎臓（kidney）
- コブクロ：子宮（uterus）
- ミノ：第一胃（rumen）の筋柱
- ハチノス：第二胃（reticulum）
- センマイ：第三胃（omasum）
- テッチャン：大腸（large intestine）

Exercises 4 の答え

問1. 1. ① b)　② d)　③ a)　④ e)　⑤ c)　2. ① c)　② a)　③ b)　3. ① d)　② b)　③ e)　④ a)　⑤ c)　4. ① f)　② d)　③ g)　④ a)　⑤ c)　⑥ e)　⑦ b)　5. ① e)　② i)　③ c)　④ d)　⑤ g)　⑥ f)　⑦ h)　⑧ b)　⑨ a)
問2. 1. ① osteitis　② osteopathy　③ osteotomy　④ osteoplasty　⑤ osteoblast
2. ① arthralgia　② arthritis　③ arthrotomy　④ arthroplasty　⑤ arthrolysis
3. ① myositis　② myology　③ myoblast　④ electromyogram　⑤ myonecrosis

Chapter 6

循環器系とその疾病
The Circulatory System and its Disorders

1 循環器系 The Circulatory System

a) 心臓 Heart

語幹	英語名詞	英語形容詞	ラテン語名詞	ギリシャ語名詞
cardi(o)-	heart	cardiac	cor	kardia

ラテン語の cor の属格（英語でいう所有格）である cordis より、英語の cordial（心からの）が派生した。

心臓にかかわる用語 Cardiac Terminology 🔊

heart	heart failure	心不全
	congestive heart failure（CHF）	うっ血性心不全
	dog heartworm	犬糸状虫
cor	cor pulmonale	肺性心
cardiac	cardiac myocyte	心筋細胞
	cardiac hypertrophy *1	心肥大
	cardiac arrest *2	心停止
cardi(o)-	myo**cardi**um, myo**cardi**a	心筋
	cardialgia, **cardio**dynia	心臓痛
	carditis	心臓炎
	cardiomyopathy	心筋症
	cardiomegaly *1	巨心症、心肥大症
	cardiopulmonary	心肺の
	cardiopulmonary resuscitation（CPR）	心肺蘇生術
	myo**cardi**tis	心筋炎
	electro**cardio**gram（ECG, EKG *3）	心電図
	phono**cardio**gram（PCG）	心音図
	echo**cardio**gram（=ultrasonography）	心エコー図

*1 cardiac hypertrophy（心肥大）は病理学的に心筋組織が肥大しているという意味で、この結果、心臓の内腔が狭まる場合と全体として広がる場合とがある。cardiomegaly（巨心症）は、臨床用語で肉眼的に心臓が大きいことを表す。すなわち cardiac hypertrophy が cardiomegaly の原因であることは多いにありうるが、cardiac hypertrophy になったからといって cardiomegaly になるとは限らない。

*2 cardiac arrest は心臓全体をひとつの器官としてとらえたとき、拍動しておらず、機能していない状態をいう。ただし、心房細胞あるいは心室細胞など個々の細胞はまだ（全体としてのリズムを作らずに）細動している可能性はある。

*3 ドイツ語の Elektrokardiogramm の頭文字。英語圏の人も EKG をよく使う。

心臓の基本形 Main Components of the Heart 🔊

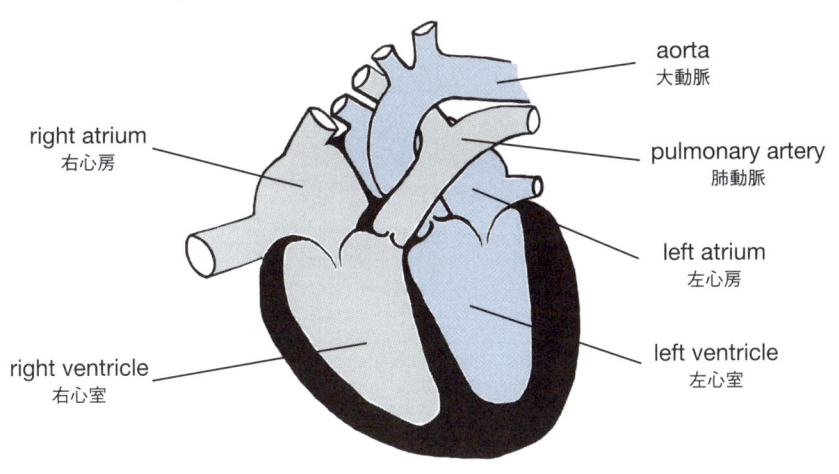

- aorta 大動脈
- pulmonary artery 肺動脈
- right atrium 右心房
- left atrium 左心房
- right ventricle 右心室
- left ventricle 左心室

心房、心室 Atria and Ventricles 🔊

atrium(sg.), atria, atriums(pl.)	心房
atrial	心房の
ventricle(sg.), ventricles(pl.)	心室
ventricular	心室の

心臓の弁 Heart Valves 🔊

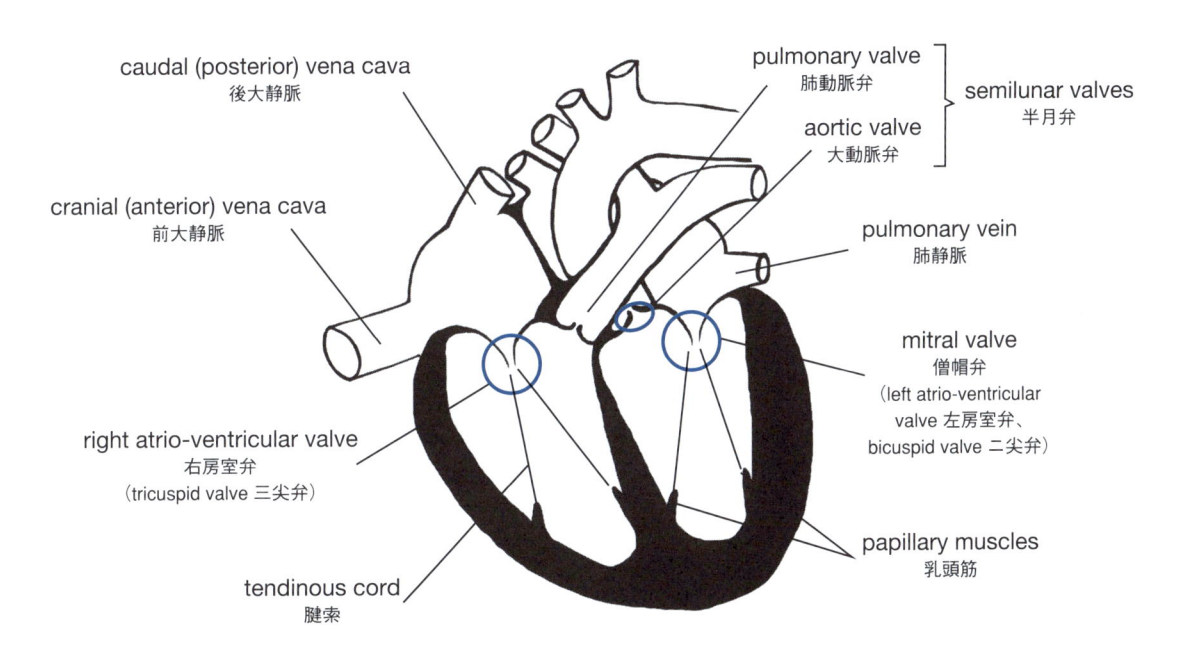

- caudal (posterior) vena cava 後大静脈
- cranial (anterior) vena cava 前大静脈
- pulmonary valve 肺動脈弁
- aortic valve 大動脈弁
- semilunar valves 半月弁
- pulmonary vein 肺静脈
- mitral valve 僧帽弁 (left atrio-ventricular valve 左房室弁、bicuspid valve 二尖弁)
- right atrio-ventricular valve 右房室弁 (tricuspid valve 三尖弁)
- papillary muscles 乳頭筋
- tendinous cord 腱索

弁の疾患 Valve Disorders 🔊 （　）内は頻発する動物種

mitral incompetence* (=mitral insufficiency, mitral regurgitation)	MR	僧帽弁閉鎖不全症 （犬）
tricuspid incompetence (=tricuspid insufficiency, tricuspid regurgitation)	TR	三尖弁閉鎖不全症 （犬、若齢猫）
pulmonary incompetence (=pulmonary insufficiency, pulmonic incompetence)	PI	肺動脈弁閉鎖不全症 （犬）
aortic incompetence (=aortic insufficiency)	AI	大動脈弁閉鎖不全症 （馬）

* 　僧帽弁閉鎖不全症はステッドマン医学大辞典には、mitral incompetence と mitral insufficiency という2つの名称が載っている。mitral（valve）regurgitation（僧帽弁逆流）の regurgitation とは、心臓の閉鎖不全の弁を通して血液が逆流することである。僧帽弁閉鎖不全と僧帽弁逆流は伴って起こるので、ほぼ同義語として使われる。
　　また、mitral incompetence の頭文字は MI である。しかし myocardial infarct（心筋梗塞）の頭文字も MI であるため、僧帽弁閉鎖不全には mitral regurgitation の頭文字の MR が用いられる。

心臓の発生に伴う疾患 Developmental Disorders*1 🔊 （　）内は頻発する動物種

patent ductus arteriosus	PDA	動脈管開存症（犬）
aortic stenosis	AS	大動脈弁狭窄
pulmonary stenosis	PS	肺動脈弁狭窄
ventricular septal defect	VSD	心室中隔欠損症（猫、馬、牛）
atrial septal defect	ASD	心房中隔欠損症
patent foramen ovale	PFO	卵円孔開存症
persistent right aortic arch	PRAA	右大動脈弓遺残（犬）*2

*1 　発生個体数は、動脈管開存症（PDA）（メスはオスの2−3倍多い）、大動脈弁狭窄（AS）、肺動脈弁狭窄（PS）の順に多い。欠損症の中でもっとも頻繁にみられるのは心室中隔欠損症（VSD）である。心房中隔欠損症（ASD）は卵円孔開存症（PFO）より広い概念で、卵円孔以外の場所に欠損がおこることも含む。
*2 　とりわけ German shepherd に多く発生する。

医学分野で重要な言葉 Important Medical Terms used for Humans 🔊

myocardial infarction, coronary thrombosis（MI）	心筋梗塞*
heart attack	心臓麻痺
heart graft, transplantation of the heart	心臓移植

* 　心筋梗塞は動物ではめったに見られない。

刺激伝導系 Impulse Conducting System 🔊

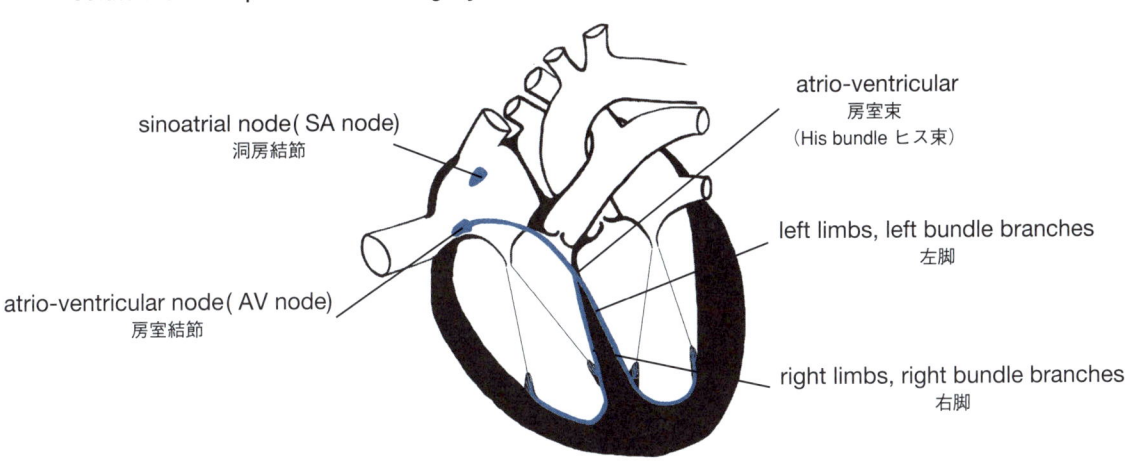

- sinoatrial node (SA node) 洞房結節
- atrio-ventricular node (AV node) 房室結節
- atrio-ventricular 房室束 (His bundle ヒス束)
- left limbs, left bundle branches 左脚
- right limbs, right bundle branches 右脚

心拍と鼓動 Heartbeat and Pulse 🔊

irregular heartbeat 異常な心拍	arrhythmia	不整脈*1
	heart block	心ブロック、ブロック*2
	bradycardia	徐脈
	tachycardia	頻脈
	premature complex, extrasystole	早期拍動、期外収縮
	atrial fibrillation	心房細動
	ventricular fibrillation	心室細動
	automated external defibrillator (AED)	自動体外式除細動器
heart murmur 心雑音	heart murmur*3	心雑音
	systole	心収縮
	systolic murmur	収縮期雑音
	asystole (→no heartbeat→death)	不全収縮*4（心音がしない、つまり死亡）
	diastole	心拡張
	diastolic murmur	拡張期雑音
	continuous murmur, machinery murmur	連続性雑音*5（機械性雑音）
auscultation		聴診
stethoscope		聴診器

*1 arrhythmia = a + rhythm + ia。リズムが狂う病気の意。
*2 心臓における刺激伝導障害で、この結果、徐脈がおこる。
*3 murmur は風のざわめき、ささやきなどという意味もある。ここでは、聴診器 (stethoscope) で聞こえる心臓などの雑音のこと。
*4 心電図などを取っていて、波形が止まり、平坦になった時をいう。
*5 連続性雑音は systolic murmur（収縮期雑音）と diastolic murmur（拡張期雑音）の両方がおこるため連続的で機械的にざーざーと聞こえる。動脈管開存のときなどにこのような音になる。continuous murmur と mechinery mumur とは、同義語として使われている。

期外収縮 Premature Complex 🔊

atrial premature complex	APC	心房性期外収縮*1
ventricular premature complex	VPC	心室性期外収縮*2

*1 SA node（洞房結節）のリズム以外に、心房で自発的に収縮がおこること。
*2 SA nodeのリズム以外に、心室で自発的に収縮がおこること。

頻脈 Tachycardia 🔊

supraventricular tachycardia	SVT	上室性頻脈（頻拍）*
ventricular tachycardia	VT	心室性頻脈（頻拍）

* 心室性でないものをすべて含む。

> **Column 田原の結節**
>
> 日本の教科書には、AV node（房室結節）のところに、しばしば「田原の結節」と書かれてある。しかし、アメリカ人心臓学者の Dr. Green は、田原という言葉を「聞いたこともない、見たこともない」そうだ。専門家さえ知らないなら、アメリカのたいていの獣医師は知らないだろう。田原淳氏は国外ではあまり知られていないのかもしれない。

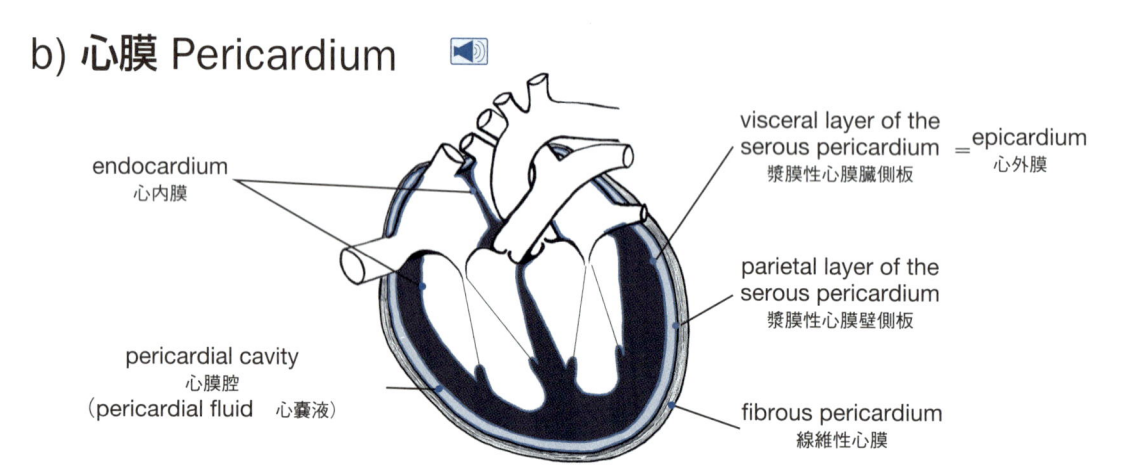

b) 心膜 Pericardium 🔊

- endocardium 心内膜
- pericardial cavity 心膜腔（pericardial fluid 心嚢液）
- visceral layer of the serous pericardium 漿膜性心膜臓側板 = epicardium 心外膜
- parietal layer of the serous pericardium 漿膜性心膜壁側板
- fibrous pericardium 線維性心膜

心膜 Pericardium 🔊

pericarditis	心膜炎（心嚢炎）
pericardial effusion	心膜液貯留、心嚢内浸出液
hydropericardium	心膜水腫
endocardium	心内膜
endocarditis	心内膜炎
pericardial fluid*	心嚢液

* 心嚢液（心膜液）は健康な状態だとやや黄色みがかった透明な液である。しかし、心膜炎などがおこると、たくさんの細胞がこの液に入るため液はにごる。また、癌などでも同じく、液の色や透明度が変わる。出血があると赤みがかる。

c) 血管 Blood Vessel

脈管 Vessel

語幹	英語名詞	英語形容詞	ラテン語名詞	ギリシャ語名詞
vas-, vaso-, vasculo-, angi(o)-	vessel	vascular	vasa sanguinea	angeion

vessel や capillary は脈管系全般を意味するため、血管だけでなくリンパ管なども含む語である。

脈管および毛細管にかかわる用語 Terminology of the Vessel and Capillary 🔊

vascular	intravascular hemolysis	血管内溶血
vasculo-	**vascul**itis（=**angi**tis）	血管炎
angi(o)-	mes**angi**um	メサンギウム、糸球体間質
	angina	アンギナ*1
	angina pectoris	狭心症
	angioma	血管腫
	angiography	血管造影
	angiogenesis	血管新生
	angiosarcoma	血管肉腫
	hem**angio**pericytoma	血管周囲腫
	hem**angio**sarcoma	血管肉腫
vessel	blood vessel	血管
	lymphatic vessel	リンパ管
capillary*2	blood capillary	毛細血管
	lymph capillary	毛細リンパ管

*1 狭心症によるしばしば絞扼感を伴う激痛のこと。
*2 英語の capillary はラテン語の caput（頭）+ pilus（毛）より。英語の capillus も頭毛のこと。毛から管を表す言葉が出たと思うと面白い。なお毛細管の直径は約8ミクロン、頭毛の直径の約1／10である。

Column 動脈は空気の通り道？

　　動脈（artery）はギリシャ語の artēria に由来するが、これは aer（空気）+ tereo（保持する）で、もともとは artery が空気の通り道だと誤解されていたことによる。
　　動脈は壁が厚く、弾性線維を多量に含むが、静脈は壁が薄く、弾力に乏しいため容易に拡張する。心臓が停止すると、心臓からの血液の拍出の圧力がなくなるので、動脈はぐっと絞まり、大部分の血液を静脈の方へと追い出してしまう。このため、古代ギリシャの人々が死体を解剖した時、動脈の中はほとんど空だった。そこで、古代ギリシャ人は動脈は空気を、静脈は血液を全身に運ぶ通路だと考え、動脈にartery と名付けてしまったのである。

d) 動脈と静脈 Arteries and Veins 🔊

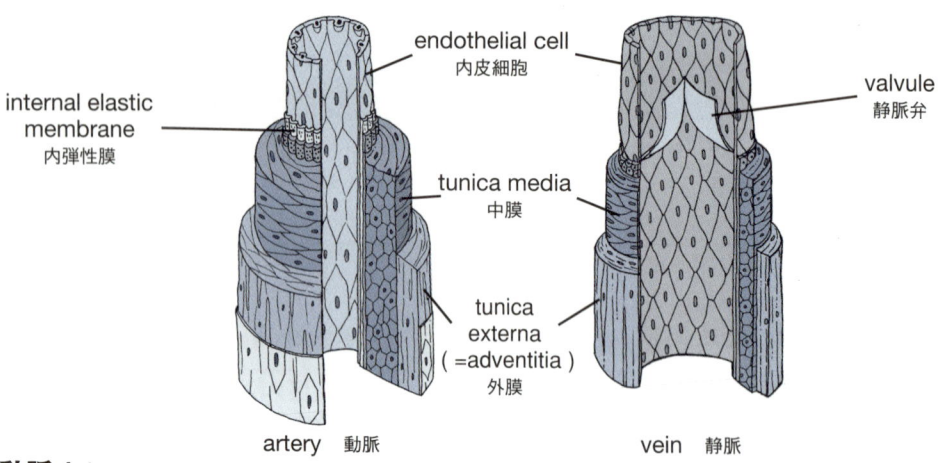

図中ラベル:
- internal elastic membrane 内弾性膜
- endothelial cell 内皮細胞
- tunica media 中膜
- tunica externa (=adventitia) 外膜
- valvule 静脈弁
- artery 動脈
- vein 静脈

動脈 Artery

語幹	英語名詞	英語形容詞	ラテン語名詞	ギリシャ語名詞
arteri(o)-	artery	arterial	arteria	artēria

静脈 Vein

語幹	英語名詞	英語形容詞	ラテン語名詞	ギリシャ語名詞
ven(o)-, veni-, vene	vein	venous	vena	-

動脈および静脈にかかわる用語 Terminology of the Artery and Vein 🔊

arteri(o)-	**arterio**sclerosis	動脈硬化症
	arteriole	小動脈
ven(o)-	**veno**stasis	静脈血うっ滞
	venule	小静脈
その他	circulatory failure	循環不全
	anastomosis	吻合
	thrombosis	血栓症
	embolism	塞栓症
	cyanosis	チアノーゼ
	aneurysm	動脈瘤
	portsystemic shunt（PSS）	門脈体循環シャント
	congestion*	うっ血、充血
	obstruction	閉塞（閉鎖）

* 医学用語ではうっ血、充血を意味するが、一般用語では、過密、渋滞を表す。例えば congestion city（過密都市）、traffic congestion（交通渋滞）である。過密になって流れが滞っている状態、それが congestion である。

主な動脈 Main Arteries 🔊

aorta	大動脈
coronary artery	冠状動脈
common carotid artery	総頸動脈

主な静脈 Main Veins 🔊

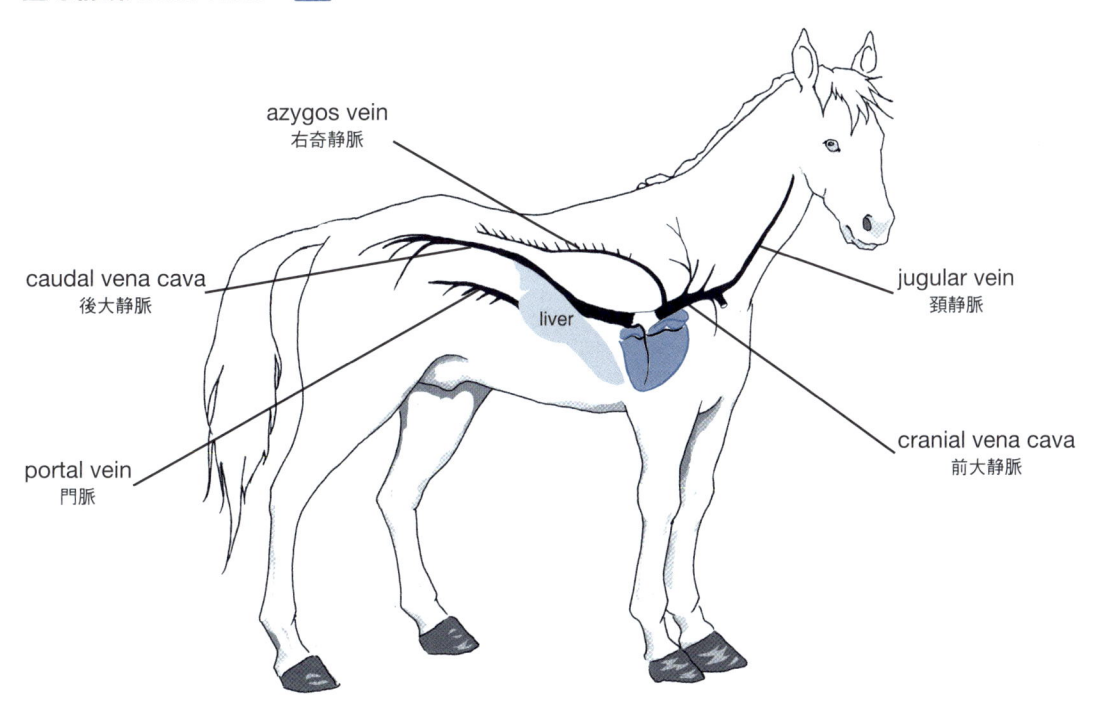

- azygos vein 右奇静脈
- caudal vena cava 後大静脈
- portal vein 門脈
- liver
- jugular vein 頸静脈
- cranial vena cava 前大静脈

Column 奇妙な静脈

奇静脈（azygos vein）は肋間の血液を集める。その名称はギリシャ語の a（〜でない）＋ zygon（対）に由来し、「対をなさない静脈」の意味である。人体では右胸腔にある方が太く発達するので、奇静脈（azygos vein）と呼び、発達の悪い左側の方は、半奇静脈（hemiazygos vein）と、それこそハンパな名前で呼ばれる。

馬、犬、猫、兎などは人間と同じタイプであり、右の azygos vein がよく発達するが、反すう類や豚では逆に左が発達するのが正常な姿である。そこで獣医解剖学では奇静脈、半奇静脈、という名称を採用せず、右奇静脈（right azygos vein）、左奇静脈（left azygos vein）と呼ぶことになっている。

メドゥーサの頭

ギリシャ神話の Medusa（メドゥーサ）は美しい髪をした女性だったが、「知恵と戦の女神アテネより私の髪の方がきれいだわ」と自慢してアテネの怒りを買い、その髪の毛、一本一本をすべて蛇に変えられてしまう。その姿を見たものはあまりのおぞましさに、石に変わってしまったという。

肝硬変などで門脈の血流が滞ると、代わりにへその周りの静脈が発達して側副路を作る。へその周囲の皮膚表面近くに、蛇がのたうつように何本もの静脈が走ることから、これを"メドゥーサの頭（Medusa head）"と呼ぶ。

e) 血液 Blood

語幹	英語名詞	英語形容詞	ラテン語名詞	ギリシャ語名詞
sangui-[*1], **hem(o)-**, **hemato-**	blood	hemal, hemic, hematic	sanguis	hāima, hāimatos[*2]

*1 語幹 sangui- は医学用語のみならず、一般用語の中でもよく使われている。sanguinary（血なまぐさい）、sanguine（血色の良い、陽気な）、consanguinity（血族）などである。
*2 hāima の属格（英語でいう所有格）。

血液にかかわる用語 Blood Terminology 🔊

blood	blood coagulation	血液凝固
	blood smear	血液塗抹
	blood transfusion	輸血
sangui-	**sangui**ferous（＝circulatory）	血液運搬の
	sanguineous	血液性の、循環血液過多の、多血質の
hem(o)-, hemato-	**hemo**globin	ヘモグロビン
	hematocrit	ヘマトクリット
	hemolymph node（＝hemal lymph node）	血リンパ節[*1]
	hemagglutination inhibition test	赤血球凝集抑制試験
	hemolytic anemia	溶血性貧血
	hematouria	血尿
	hematothorax, **hemo**thorax	血胸
	hemorrhagic diseases	出血性疾患
	hemolysis	溶血
	hemophilia	血友病
-emia[*2]	an**emia**[*3]	貧血
	septic**emia**	敗血症[*4]
	isch**emia**	虚血
	leuk**emia**	白血病
	bacter**emia**	菌血症
	vir**emia**	ウイルス血症
	hyper**emia**	充血
	hypercalc**emia**	高カルシウム血症
	chyla**emia**	乳び血症
その他	scurvy	壊血病
	iron deficiency	鉄欠乏
	hypoxia	低酸素症
	cyanosis	チアノーゼ
	congestion	うっ血

*1 反すう類のみに見られる。
*2 血液の病気を表す接尾辞に -emia がある。ギリシャ語の「haima（血液）」から"h"が取れ、残りの aima が変形したものである。
*3 a- は、a- 以下の言葉を打ち消す用語。anemia は血を打ち消しているので「貧血」という意味になる。
*4 septic：腐敗性の、敗血性の。

血液とリンパ液の流れ Circulation of Blood and Lymph 🔊

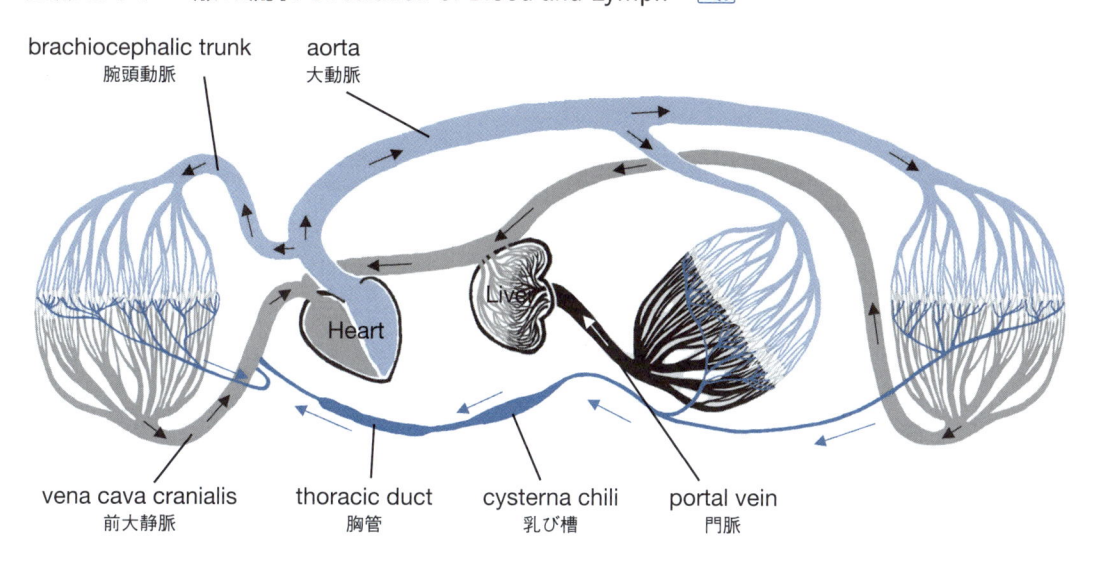

- brachiocephalic trunk 腕頭動脈
- aorta 大動脈
- Heart
- Liver
- vena cava cranialis 前大静脈
- thoracic duct 胸管
- cysterna chili 乳び槽
- portal vein 門脈

　動脈血は毛細血管を経て、その約85％は静脈に至る。消化管からの静脈血は、門脈を通って肝臓に入り、肝臓内で再度、毛細血管となる。そののち集まって、後大静脈に加わる（肺循環は省略）。また、血漿成分の約15％は毛細リンパ管へ入り、静脈とは別の流れを作る。
　リンパ液は最終的には前大静脈に加わり、静脈血と混ざって心臓へ戻る。

血圧と血糖　Blood Pressure and Blood Glucose 🔊

high blood pressure （=hypertension）	高血圧
low blood pressure （=hypotension）	低血圧
blood sugar level （=blood glucose level）	血糖値
high blood glucose level （=hyperglycaemia, hyperglycemia）	高血糖
low blood glucose level （=hypoglycaemia, hypoglycemia）	低血糖

出血 Blood Loss 🔊

bleeding	出血すること
internal bleeding	内出血
nasal bleeding, epistaxis[*1]	鼻血
hemorrhage[*2]	出血

*1　nasal bleeding は一般用語、epistaxisは医学用語である。
*2　hemo（血）+ rrhage（ギリシャ語の rhegnymi：突発する。飛び出す）より、出血、とりわけ突然の大出血（血管壁が破れて内出血するなど）ことを言う。bleeding は一般用語で、出血の種類にかかわらず用いられるが、hemorrhage は医学用語的イメージが強い。

f) 血球 Blood Cells

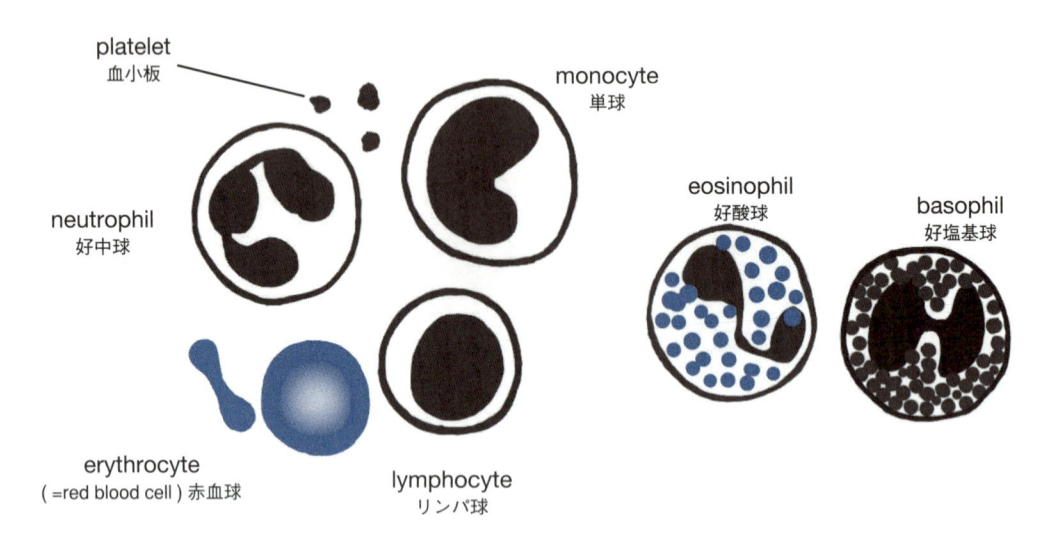

platelet 血小板
monocyte 単球
neutrophil 好中球
eosinophil 好酸球
basophil 好塩基球
erythrocyte (=red blood cell) 赤血球
lymphocyte リンパ球

血球 Blood Cells

erythrocyte, red blood cell（RBC）, red corpuscle			赤血球
leukocyte, white blood cell （WBC）, white corpuscle 白血球	granulocyte, granular leukocyte 顆粒球	neutrophil, neutrophile, neutrophilic leukocyte	好中球
		eosinophil, eosinophile, eosinocyte, eosinophilic leukocyte, acidophilic leukocyte, oxyphil, oxyphile, oxyphilic leukocyte	好酸球
		basophil, basophile, basocyte, basophilic leukocyte	好塩基球
	lymphocyte, lymph corpuscle, lymphatic corpuscle		リンパ球
	monocyte		単球
platelet, thrombocyte			血小板

血漿と血清 Plasma and Serum

plasma*1		血漿
serum*2 血清	acute stage serum	急性期血清
	convalescent stage serum	回復期血清

*1 循環血液中の液性成分。血球は含まないが、フィブリンは含む。
*2 凝血後の血液の液性成分。フィブリンと血球を除いたもの。

血液細胞の疾病 Blood Cell Disorders

leukocyte 白血球	leucopenia／leukopenia	白血球減少症
	feline panleukopenia	猫汎白血球減少症
	feline leukemia	猫白血病
	leukocyte adhesion deficiency	白血球粘着不全症
lymphocyte リンパ球	lymphoma	リンパ腫
mast cell 肥満細胞	mast cell tumor（MCT）	肥満細胞腫
monocyte 単球	monocytosis	単球増加症
histocyte, histiocyte 組織球	histiocytoma	組織球腫
	malignant fibrous histiocytoma	悪性線維性組織球腫

骨髄 Bone Marrow

hematopoiesis, hemopoiesis, hematogenesis, blood formation	造血
hematopoietic tissue, blood-forming tissue	造血組織
megakaryocyte	巨核球
myeloid tissue	骨髄組織

Column 勇気と桜

　ラテン語で心臓は"Cor"。ここからフランス語の心臓 coeur が起こり、これが英語となり、courage（勇気）が派生した。心臓と勇気、なんとなくぴったりくる。Cor からは、英語の core（核）も派生した。

　英語の「血液」は blood、「出血する」は bleed であり、「花咲く」の bloom と同じ語源で、祖インド・ヨーロッパ語の bhle- に由来する。何かが裂けて飛び出してくることを言い、蕾が裂けて、花弁が飛び出せば花が咲き、皮膚や内臓が裂けて、血が飛び出せば出血なのである。これを学んだ時、日本語の「咲く」との類似に驚いた。日本語の「咲く」は「裂く」に由来するとされ、蕾が裂けて花が開くことを言う。つまり英語の bloom と全く同じアイデアなのだ。ちなみに桜は「咲く」に接尾辞の「ら」が付いたものである。"咲くもの"それが桜なのである。

2 リンパ器官 Lymphatic Organs

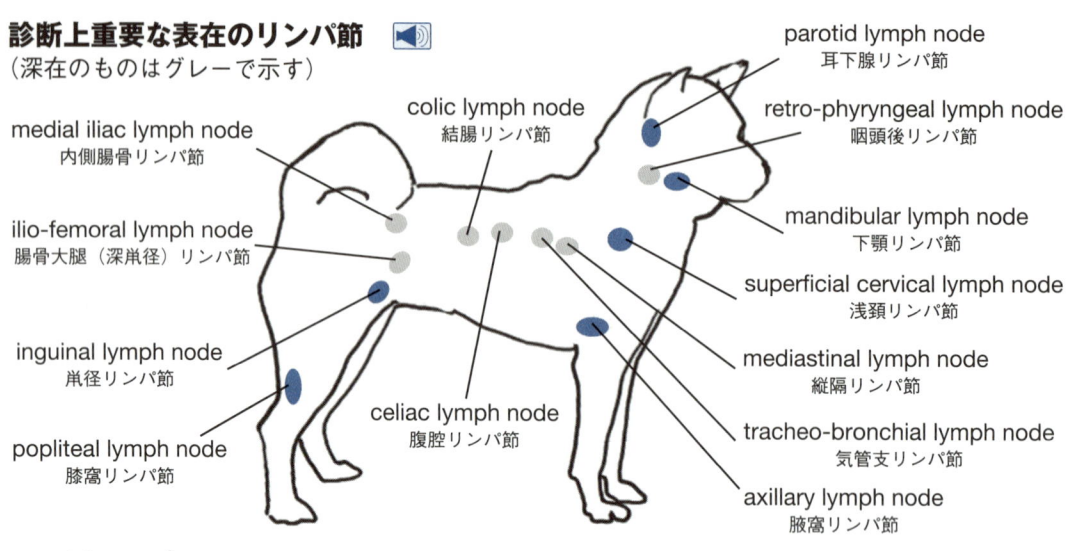

診断上重要な表在のリンパ節
(深在のものはグレーで示す)

- medial iliac lymph node 内側腸骨リンパ節
- ilio-femoral lymph node 腸骨大腿（深鼠径）リンパ節
- inguinal lymph node 鼠径リンパ節
- popliteal lymph node 膝窩リンパ節
- colic lymph node 結腸リンパ節
- celiac lymph node 腹腔リンパ節
- parotid lymph node 耳下腺リンパ節
- retro-phyryngeal lymph node 咽頭後リンパ節
- mandibular lymph node 下顎リンパ節
- superficial cervical lymph node 浅頚リンパ節
- mediastinal lymph node 縦隔リンパ節
- tracheo-bronchial lymph node 気管支リンパ節
- axillary lymph node 腋窩リンパ節

a) リンパ Lymph

語幹	英語名詞	英語形容詞	ラテン語名詞	ギリシャ語名詞
lymph(o)-	lymph	lymphatic	lympha	-

リンパにかかわる用語 Lymphatic Terminology

lymph(o)-	**lympho**cyte		リンパ球
	lymphonode, lymph node, **lymph**aden		リンパ節
	lymphoma		リンパ腫
	lymphosarcoma		リンパ肉腫
lymph	lymph nodule リンパ小節	solitary lymphatic nodule	孤立リンパ小節
		aggregated lymphatic nodule	集合リンパ小節
		Peyer's patch	パイエル板
	lymph vessel リンパ管	lymphatic vessel, lymph vessel, lymphoduct	リンパ管
		thoracic duct	胸管
		cisterna chyli*1	乳び槽
		lymph capillary	毛細リンパ管
lymphaden(o)-*2	**lymphaden**itis		リンパ節炎
	lymphadenosis		リンパ節症
	lymphadenectomy		リンパ節切除、リンパ節廓清
	lymphadenopathy		リンパ腺症、リンパ腺病*3
lymphangi-*4	**lymphangi**tis		リンパ管炎
	lymphangiectasis		腸リンパ管拡張症
	lymphangioma		リンパ管腫

*1 chyl(o)-（乳び）は、ギリシャ語の chylos（果物や植物の汁）に由来する。chylo- を含む用語には chylemia（乳び血症）、chylorrhea（乳び漏）などがある。
*2 リンパ節を意味する連結形。lympho- + aden（ギリシャ語の腺）より。
*3 リンパ節の慢性的腫大。
*4 リンパ管に関する接続形。lymph- + angi（管）。

b) リンパ系器官　Other Lymphatic Organs

扁桃 Tonsil

語幹	英語名詞	英語形容詞	ラテン語名詞	ギリシャ語名詞
tonsill(o)-	tonsil	tonsillar, tonsillary	tonsilla(sg.), tonsillae(pl.)	-

胸腺 Thymus

語幹	英語名詞	英語形容詞	ラテン語名詞	ギリシャ語名詞
thym(o)-	thymus(sg.), thymi, thymuses(pl.)	thymic	thymus	thymos

ファブリキウス嚢 Bursa of Fabricius*

語幹	英語名詞	英語形容詞	ラテン語名詞	ギリシャ語名詞
-	bursa of Fabricius	bursal	bursa cloacalis	-

* bursa of Fabriciusは鳥類が持つ。

リンパ系器官にかかわる用語 Terminology of the other Lymphatic Organs

語幹	英語	日本語
tonsil	palatine tonsil	口蓋扁桃
tonsill(o)-	**tonsill**itis	扁桃炎
	tonsillectomy and adenoidectomy	扁桃・アデノイド摘出術
thym(o)-	**thymo**ma	胸腺腫

> **Column　泉の精霊**
>
> 　リンパはラテン語の lympha（清らかな泉の水）に由来する。ギリシャの精霊、"ニンフ（Nymphe）" の N が L に置き換わった、という説がある。ニンフは若く美しい女性の姿をしていて、歌と踊りを好み、庭園や牧場に花を咲かせたり、家畜を見張ったり、狩りの獲物を提供したり、清らかな泉や湖を守ったりするのだという。
>
> **タイムとthymus**
>
> 　タイム（thyme）はいわゆるハーブ（香草）の一種で、肉料理やジャガイモ料理、タイム・ティーなど、におい消しや香り付けとして用いられる。和名の立麝香草（たちじゃこうそう）の漢字からも、香がたちのぼってくるようだ。
> 　タイムの学名は *Thymus vulgaris* である。このハーブの名前は、胸腺（thymus）の起源となった。Oxford辞書には胸腺のことを「タイムの芽のようないぼ状であるから」とある。タイムは小さな丸い花の芽が枝の先に多数集まって、全体として丸まった形になる。古代ギリシャの人々は胸腺が小葉の集合であるのを見て、タイムの芽を連想したのかもしれない。

C) 脾臓 Spleen

語幹	英語名詞	英語形容詞	ラテン語名詞	ギリシャ語名詞
splen(o)-	spleen	splenic	lien	splen

脾臓にかかわる用語 Spleen Terminology 🔊

splen(o)-	**splen**itis	脾炎
	traumatic **splen**itis	外傷性（創傷性）脾炎
	splenomegaly	脾腫（巨脾症）
	splenectomy	脾臓切除術、脾摘出
	splenectopia, **splen**ectopy	脾転位症
その他	red pulp	赤脾髄
	white pulp	白脾髄

Column かんしゃくの座

英語の spleen には、脾臓だけでなく、「怒り」や「かんしゃく」という意味もある。古代ギリシャでは脾臓に不機嫌、かんしゃくの座があると考えられていたことによるものだ。"He vented his spleen." と言うと、「彼は怒りをぶちまけた」という意味である。

免疫

免疫性（immunity）はラテン語の immunitas に由来する。im- は「〜でない」という否定、munis は「公の義務」という意味で、「公の義務を免ぜられる」からそのようなイヤな病気を免ぜられるへと転じて「免疫」となった。同じ munis を含む語に community がある。con- は「〜と共に」を意味し、「公の義務」を共にすることから、「連帯感、共同社会」となった。

Immunity に話を戻すと、immunity from taxationといえば課税を免じられている、つまり免税、である。diplomatic immunity は「外交官の特権」のことで、ふつうの人には課せられる面倒な義務だが、外交官には「免じられている」ものを言う。

Immunity とは公の義務であれ、病気であれ、税金であれ、うれしくないものを免ぜられることなのである。

3 免疫学 Immunology

免疫にかかわる用語 Immunological Terminology

免疫は語幹"immun(o)-"を用いて表される。

immune	immune response	免疫応答
immun(o)-	**immun**ization	免疫
	immunogenicity	免疫原性
	immunoenhancement	免疫増強
	immunoprecipitation	免疫沈降
	immunodeficiency	免疫不全
	immunosuppression	免疫抑制
	immunohistochemistry	免疫組織化学
	immunofluorescence	免疫蛍光検査法

細網内皮系 Reticuloendothelial System（RES）

cellular immunity	細胞性免疫
humoral immunity	液性免疫

免疫応答に関わる細胞 Cells Related to Immunoresponses

B cell	B細胞（Bリンパ球）*1
T cell	T細胞（Tリンパ球）*2
plasma cell, plasmacyte	形質細胞
mast cell, mastocyte	肥満細胞
neutrophil	好中球
dendritic cell	樹状細胞
macrophage	マクロファージ、大食細胞
phagocytosis／phagocytize, phagocytose	貪食／貪食する

＊1　bone marrow（骨髄）の頭文字に由来。
＊2　thymus（胸腺）の頭文字に由来。

抗原、抗体、補体 Antigen, Antibody and Complement

antigen	抗原
antibody	抗体
antigen-antibody reaction	抗原抗体反応
antigen presenting cell（APC）	抗原提示細胞
immunoglobulin	免疫グロブリン
complement	補体
complement fixation	補体結合
cytokine	サイトカイン

免疫不全 Immunodeficiency

acquired immune deficiency syndrome (AIDS)	後天性免疫不全症候群（エイズ）
human immunodeficiency virus (HIV)	ヒト免疫不全ウイルス
opportunistic infection	日和見感染
Kaposi sarcoma	カポシ肉腫
Pneumocystis carinii pneumonia	カリニ肺炎
feline immunodeficiency virus (FIV)	猫免疫不全ウイルス
feline acquired immunodeficiency syndrome	猫後天性免疫不全症候群（猫エイズ）
combined immunodeficiency	複合免疫不全症

自己免疫疾患 Autoimmune Diseases

autoimmune disease	自己免疫疾患
collagenosis, collagen diseases, collagen-vascular diseases	膠原病
articular rheumatism, rheumatoid arthritis	関節リウマチ
systemic lupus erythematosus (SLE)	全身性エリテマトーデス
multiple sclerosis (MS)	多発性硬化症

アレルギー Allergy

allergy		アレルギー
allergen		アレルゲン
atopy アトピー	atopic dermatitis	アトピー性皮膚炎
	atopic asthma	アトピー性喘息
anaphylaxis		アナフィラキシー
anaphylactic shock		アナフィラキシーショック

ワクチン、輸血 Vaccine and Transfusion

vaccine	ワクチン
infusion	輸液
blood transfusion	輸血

移植 Transplantation

major histocompatibility complex (MHC)	主要組織適合遺伝子複合体
transplantation	移植
graft	移植片、グラフト
rejection	拒絶（反応）

問1 次の日本語を英語に直しなさい。

① 心臓 _____
② 心臓の _____
③ 心膜 _____
④ 心膜炎 _____

問2 下の語群から（　）に適当なものを選びなさい。

1. 心房（①　　　）と心室（②　　　）の間の房室弁（③　　　）は左心の二尖弁、右心の三尖弁であり、英語もそれぞれ、2と3を表す接頭辞を付けて、（④　　　）と、（⑤　　　）となる。臨床分野では左房室弁は、僧帽弁（⑥　　　）という名称が頻用される。

    a) mitral valve　b) bicuspid valve　c) tricuspid valve
    d) atrioventricular valves　e) atrium　f) ventricle

2. 胎子期、ガス交換は肺ではなく、胎盤（①　　　）で行われる。このため、胎子には、肺動脈と大動脈を結ぶライン、すなわち動脈管（②　　　）がある。通常、出生前後に閉じるが、これが閉鎖せず開いたままになってしまうのが動脈管開存症（③　　　）である。
   また、先天性心疾患の集合に、肺動脈弁狭窄、大動脈の右室騎乗（大動脈が動脈血だけでなく静脈血も受ける）、心室中隔欠損および右心室肥大の四大徴候、ファロー四徴症（④　　　）がある。

    a) arterial canal（ductus arteriosus）　b) tetralogy of Fallot
    c) placenta　d) patent ductus arteriosus

問3 貧血にかかわる代表的な用語を挙げた。左右を正しく結びなさい。

① 溶血性貧血　　　　・　　・ a) blood loss anemia
② 失血性貧血　　　　・　　・ b) dyserythropoietic anemia
③ 赤血球異形成貧血　・　　・ c) hemolytic anemia

問4 溶血性貧血を起こす病名の左右を正しく結びなさい。

① babesiosis　　　　　　　　　・　　・ a) 再生不良性貧血
② theileriasis　　　　　　　　　・　　・ b) 馬伝染性貧血
③ anemia of renal failure　　　・　　・ c) バベシア症
④ equine infectious anemia　　・　　・ d) 慢性腎炎に伴う貧血（腎性貧血）
⑤ isoimmune hemolytic anemia ・　　・ e) タマネギ中毒
⑥ onion poisoning　　　　　　　・　　・ f) タイレリア症
⑦ aplastic anemia　　　　　　　・　　・ g) 鉄欠乏性貧血
⑧ iron deficiency anemia　　　・　　・ h) 新生子黄疸

問5．白血球にかかわる病名の左右を正しく結びなさい。

①白血病　　　　　　　　　・　　・ a) leukopenia
②リンパ腫　　　　　　　　・　　・ b) leukemia
③白血球減少症　　　　　　・　　・ c) feline immuno deficiency virus infection
④猫免疫不全ウイルス感染症・　　・ d) multiple myeloma
⑤多発性骨髄炎　　　　　　・　　・ e) lymphoma

問6．下の語群から適切な用語を選びなさい。

1. 血栓症 _____
2. 塞栓症 _____
3. 小血管 _____
4. 動脈瘤 _____
5. 犬糸状虫症 _____
6. 肺動脈 _____

small vessels,　pulmonary artery,　embolism,　dirofilariasis,　thrombosis,　aneurysm,

Column　キリンの首の秘密

　キリンは長い首を通して、高い位置にある脳に血液を送らなくてはならないからだろう、とんでもなく高血圧な動物である。正常な動脈血圧は260／160mmHgもある。人間では、血圧が140／90mmHgだと軽度ないし中度の高血圧、180／110mmHgを超えると重度の高血圧だが、キリンの正常値は、これらを軽々と超えているのである。

　キリンが川辺で水を飲む時、脚を踏ん張り、高い位置にある頭を水辺まで下げなくてはいけない。私たちも逆立ちして頭を下げたら、頭部に血液がなだれ込んで来るのを感じる。ただでさえ高血圧なキリンは、急に頭を下げたらくらくらしないだろうか？

　実はキリンを含む偶蹄類（artiodactyla）には、脳下部に怪網（rete mirabile）と呼ばれる、その名も怪しい中動脈の網状構造が発達している。水を飲む時、この怪網に過剰な血液が流れ込むことにより、「頭に血が登った」状態のバッファーとして働くらしい。怪網は反すう類や豚にもある。牛や豚での怪網の役割はよく分かっていないが、進化の過程上、これをうまく利用することで、キリンの祖先は首を高く伸ばすことに成功したのかもしれない。

Exercises 5 の答え

問1．① heart　② cardiac　③ pericardium　④ pericarditis
問2．1. ① e)　② f)　③ d)　④ b)　⑤ c)　⑥ a)　　2. ① c)　② a)　③ d)　④ b)
問3．① c)　② a)　③ b)
問4．① c)　② f)　③ d)　④ b)　⑤ h)　⑥ e)　⑦ a)　⑧ g)
問5．① b)　② e)　③ a)　④ c)　⑤ d)
問6．① thrombosis　② embolism　③ small vessels　④ aneurysm　⑤ dirofilariasis　⑥ pulmonary artery

Chapter 7

呼吸器系とその疾患
The Respiratory System and its Disorders

1 呼吸器系 The Respiratory System

a) 呼吸 Respiration

語幹	英語名詞	英語形容詞	ラテン語名詞	ギリシャ語名詞
pneo-, pneum-, -pnea	respiration	respiratory	respiratio	pneo

　呼吸は医学用語では respiration 、一般用語では breathing で表される。ラテン語の respiratio は re-（再び）+ spirare（息をする）。また、inspiration（吸息）は in-（内へ）+ spirare 、expiration（呼息）は ex-（外へ）+ spirare である。
　pneumo- は空気、呼吸または肺を意味する。

呼吸にかかわる用語 Respiratory Terminology 🔊

pneo-, pneum-	**pneo**cardiac reflex	吸入性心臓反射
	pneumatosis	気腫、気症
-pnea	a**pnea**[*1]	無呼吸
	dys**pnea**[*2]	呼吸困難
respiration	abdominal respiration （=abdominal breathing）	腹式呼吸
respiratory	respiratory acidosis	呼吸性アシドーシス

＊1　a-（否定）+ -pnea 。
＊2　dys-（悪化、困難）+ -pnea 。

b) 鼻腔 Nasal Cavity

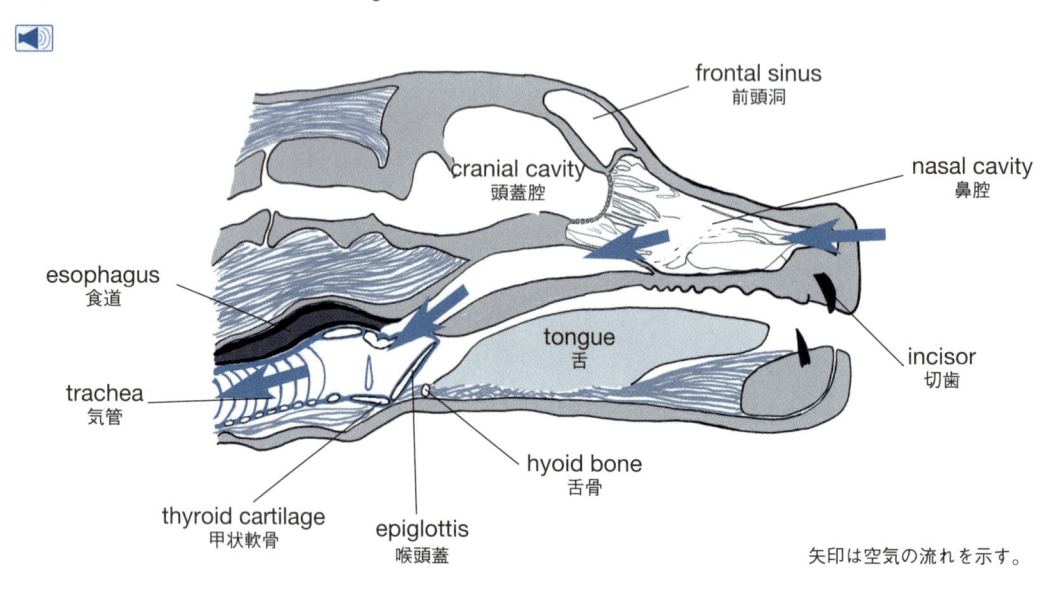

🔊

frontal sinus 前頭洞
cranial cavity 頭蓋腔
nasal cavity 鼻腔
esophagus 食道
tongue 舌
incisor 切歯
trachea 気管
hyoid bone 舌骨
thyroid cartilage 甲状軟骨
epiglottis 喉頭蓋

矢印は空気の流れを示す。

鼻腔にかかわる用語 Terminology of the Nasal Cavity 🔊

rhin-*	**rhin**itis		鼻炎
	swine atrophic **rhin**itis（AR）		萎縮性鼻炎
その他	infectious coryza（acute **rhin**itis）		伝染性コリーザ
	external nose 外鼻	muzzle	鼻づら（犬、猫）
		nostril, naris (sg.), nares (pl.)	外鼻孔
	nasal cavity 鼻腔	nasal discharge	鼻汁
		epistaxis（=nose bleeding）	鼻出血
	paranasal sinuses 副鼻腔	sinusitis（=empyema）	副鼻腔炎、蓄膿症

＊ 鼻を意味する語幹。

c) 喉頭 Larynx

語幹	英語名詞	英語形容詞	ラテン語名詞	ギリシャ語名詞
laryng(o)-	larynx	laryngeal	larynx	larynx

咽頭にかかわる用語 Larynx Terminology 🔊

laryng(o)-	**laryng**itis 喉頭炎	acute（catarrhal）**laryng**itis		急性（カタル性）喉頭炎
		chronic **laryng**itis		慢性喉頭炎
		croupous **laryng**itis		クループ性喉頭炎
laryngeal	laryngeal paralysis			喉頭麻痺
その他	epiglottis*1			喉頭蓋
	vocal cord 声帯	vocal muscle		声帯筋
		chorditis*2		声帯炎
		debarking, ventriculocordectomy		喉頭声帯除去術（犬）
	roaring, laryngeal paralysis, laryngohemiplegia			喘鳴症または喉頭（片）麻痺（馬）
	snore いびき*3	stertorous respiration		鼾声呼吸
		himantosis		口蓋垂延長

＊1　epi-（上）+ glottis（声門）より。形容詞は epiglottic , epiglottidean 。
＊2　chord の炎症を表す用語のため、spinal cord（脊髄）や spermatic cord（精索）の炎症にも chorditis を用いる。
＊3　He snored loudly.（彼は大いびきをかいた）。

d) 胸郭と胸膜 Thorax and Pleura

胸郭 Thorax

語幹	英語名詞	英語形容詞	ラテン語名詞	ギリシャ語名詞
thorac(o)-, thoracico-, -thorax	thorax (sg.) thoraxes, thoraces (pl.)	thoracic	thorax	thorax

胸膜 Pleura

語幹	英語名詞	英語形容詞	ラテン語名詞	ギリシャ語名詞
pleura-, pleur(o)-	pleura	pleural	pleura	pleura*

* 肋骨、脇の意。

胸郭および胸膜にかかわる用語 Terminology of the Thorax and Pleura 🔊

thoracic	thoracic cavity	胸腔
	thoracic duct（TD）	胸管
thorac(o)-	**thoraco**stomy tube	胸腔ドレーン
-thorax	pyo**thorax**（=empyema）*	膿胸
	hemo**thorax**	血胸
	hydro**thorax**	水胸
	chylo**thorax**	乳び胸
	pneumo**thorax**	気胸
pleur(o)-	**pleur**itis, **pleur**isy	胸膜炎
pleural	pleural effusion	胸水
その他	mediastinum	縦隔
	mediastinal emphysema, pneumomediastinum	縦隔気腫

* 膿胸と蓄膿症の2つの意味がある。

> **Column 悪い空気**
>
> マラリア（malaria）はコクシジウム目に属する原虫によって赤血球が破壊される疾病で、ハマダラ蚊により媒介される。しかし昔の人々は、"悪い空気によってもたらされる病気"と考えた。そこで2つのラテン語、mal-（悪い）と aria（空気）を合成し、malaria（マラリア）と命名した。

e) 横隔膜 Diaphragm

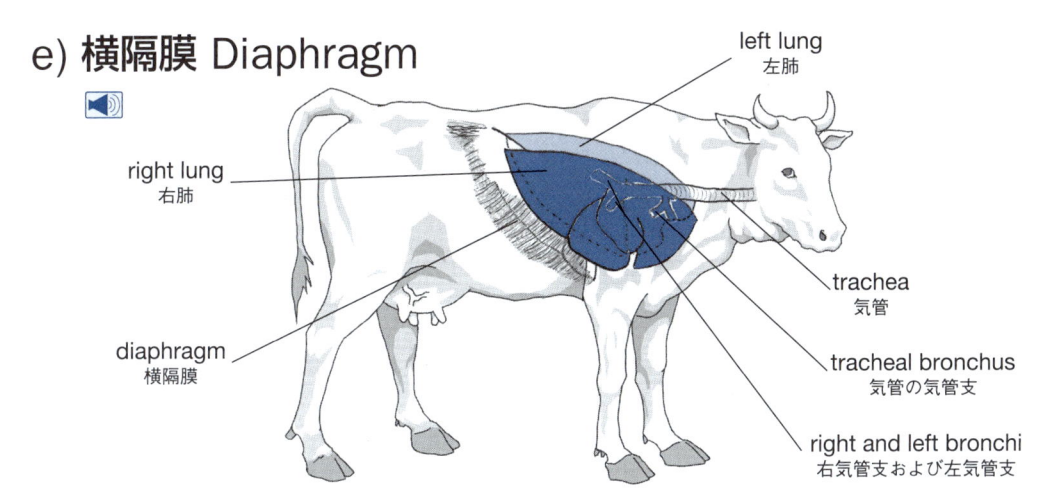

left lung 左肺
right lung 右肺
diaphragm 横隔膜
trachea 気管
tracheal bronchus 気管の気管支
right and left bronchi 右気管支および左気管支

語幹	英語名詞	英語形容詞	ラテン語名詞	ギリシャ語名詞
phren-, phreni-, phreno-, phrenico-	diaphragm	diaphragmatic, phrenic	diaphragma	diaphragma*

＊ diaphragma = dia- (〜を通して) + phragma (囲い、フェンス) に由来する。

横隔膜にかかわる用語 Diaphragm Terminology

phrenic	phrenic nerve	横隔神経
phren-	**phren**itis	横隔膜炎
diaphragmatic	diaphragmatic hernia	横隔膜ヘルニア
	synchronous diaphragmatic flutter in horses	馬の同時性横隔膜粗動
その他	thumps	吃逆(きつぎゃく)
	hiccup	しゃっくり*

＊ 横隔膜とその周辺の痙攣を指す。

> **Column 精神の座はいずこに**
>
> 横隔膜 (diaphragm) の形容詞のひとつである "phrenic" は、この名詞形とは全く違う言葉である。なぜだろうか？
>
> phrenic はギリシャ語の phren, phrenos に由来し、心、精神という意味である。古代ギリシャの人々は "精神の座が横隔膜に宿る" と考えたため、phrenic が精神と横隔膜を同時に表す言葉となった。例えば schizophrenia と言えば統合失調症（精神分裂病）である (schizo- は分裂)。
>
> phrenitis は以前は「脳炎」と「横隔膜炎」の両方の意味があったが、現在は主に横隔膜炎の意味で使われ、脳炎には encephalitis や cerebritis が使われる。

f) 気管と気管支 Trachea and Bronchi

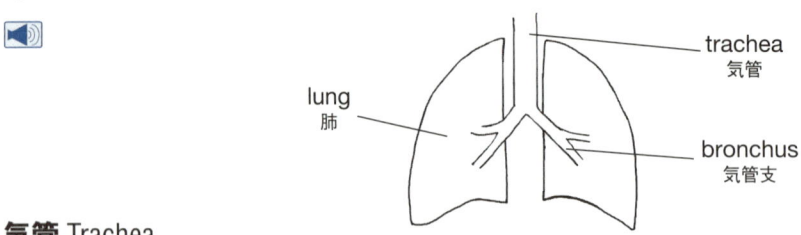

気管 Trachea

語幹	英語名詞	英語形容詞	ラテン語名詞	ギリシャ語名詞
trache(o)-	trachea(sg.), tracheae(pl.)	tracheal	trachea	tracheia

気管支 Bronchi

語幹	英語名詞	英語形容詞	ラテン語名詞	ギリシャ語名詞
bronch(o)-, bronchi-	bronchus(sg.), bronchi(pl.)	bronchial	bronchus(sg.)* bronchi(pl.)	bronchos

気管と気管支は、日本語では似ているが、英語ではまったく異なる言葉であることに注意！

気管および気管支にかかわる用語 Terminology of the Trachea and Bronchi

trache(o)-	**tracheo**stomy	気管開口［術］
	tracheitis, **trach**itis	気管炎
	tracheotomy	気管切開［術］
tracheal	tracheal collapse	気管虚脱
bronch(o)-, bronchi-	**bronch**itis	気管支炎
	bronchopneumonia	気管支肺炎
	bronchiectasis	気管支拡張症
	air **brohcho**gram	エアブロンコグラム（肺胞空気充満像）

g) 肺 Lung

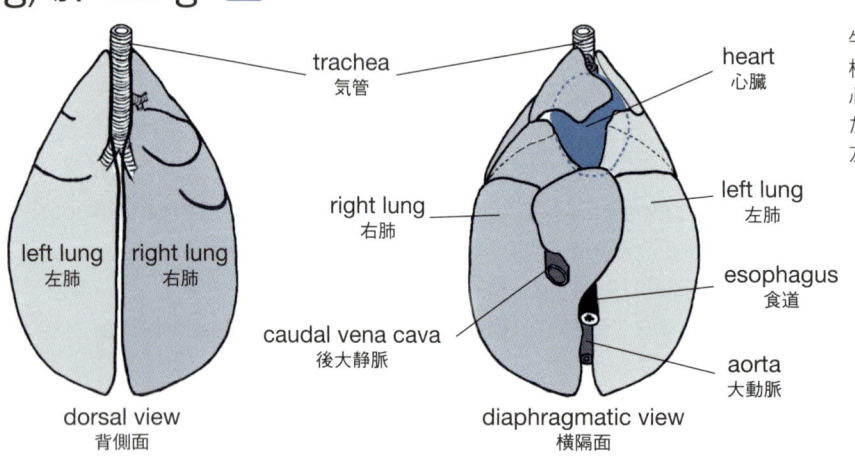

牛の肺の背側方および横隔膜面からの模式図。心臓は正中より、わずかに左に位置し、肺は左肺より右肺が大きい。

dorsal view 背側面
- trachea 気管
- left lung 左肺
- right lung 右肺

diaphragmatic view 横隔面
- heart 心臓
- right lung 右肺
- left lung 左肺
- caudal vena cava 後大静脈
- esophagus 食道
- aorta 大動脈

語幹	英語名詞	英語形容詞	ラテン語名詞	ギリシャ語名詞
pneum(o)-, pneuma-, pneumono-, pneumat(o)-	lung	pulmonary	pulmo	pneumōs

肺にかかわる用語 Lung Terminology 🔊

pulmonary	pulmonary artery		肺動脈
	pulmonary valve		肺動脈弁
	plumonary lobectomy		肺葉切除術[*1]
	pulmonary edema		肺水腫
	pulmonary tuberculosis		結核（肺結核）[*2]
	pulmonary infarct		肺梗塞症
	pulmonary emphysema		肺気腫
pneum(o)-	**pneumo**nia[*3] 肺炎	catarrhal **pneumo**nia	カタル性肺炎
		croupous **pneumo**nia	クループ性肺炎
		interstitial **peumo**nia	間質性肺炎
		aspiration **pneumo**nia	誤嚥性（吸引性）肺炎
	pneumothorax		気胸
	pneumoconiosis		塵肺
lung	lung cancer		肺癌[*4]
	lung abscess		肺膿瘍[*5]
	lungworm		肺線虫
その他	measles		はしか
	bullae		肺嚢胞症[*6]
	atelectasis		無気肺

[*1] 片側の肺をすべて除去する pneumonectomy（肺切除術）との混同に注意。
[*2] 結核菌 *Tubercle Bacillus* によっておこる。この頭文字をとって、TB（ティービー）ということあり。
[*3] 肺炎は pneumonia。pneumonitis という言葉もあるが使用頻度が低い。
[*4] 悪性。pulmonary tumor より一般的に使用される。
[*5] 肺内に膿が巣状に貯留した状態をいう。
[*6] 先天性の肺内の異常空洞が見られる。

「肺の」は pulmonary だが、「肺胞の」は alveolar である

alveolar には「肺胞の」と同時に、「歯槽の」という意味もある。例えば alveolar bone は歯槽骨である。ラテン語の alveus（空洞）に指小辞 "l" が入って「小さな空洞」を意味するようになった。

🔊

alveolus (sg.)	肺胞
alveoli (pl.)	
alveolar	肺胞の
alveolar sac	肺胞嚢

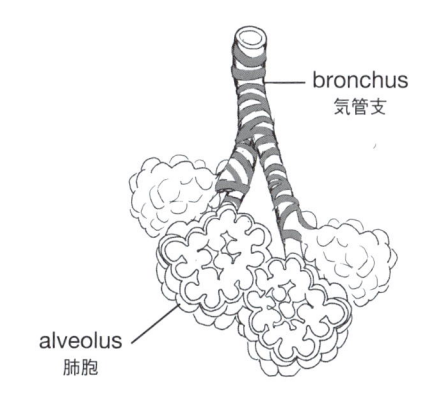

bronchus 気管支

alveolus 肺胞

Exercises 6

問1．-itis（炎症）とつないで、次の用語を作りなさい。

① 鼻炎 _____
② 喉頭炎 _____
③ 気管炎 _____
④ 胸膜炎 _____
⑤ 肺炎 _____

問2．-pnea（呼吸）とつないで、次の用語を作りなさい。

① 無呼吸 _____
② 呼吸困難 _____
③ 正常呼吸 _____
④ 呼吸低下 _____
⑤ 過呼吸 _____

問3．（ ）に下の語群から適切な用語を選んで入れなさい。

1. 胸腔内に膿性浸出液が貯留した状態を膿胸（① _____）という。py- あるいは pyo- という膿を表す接頭辞と胸腔（② _____）に分けられる。独立した単語で、膿そのものを指すには、（③ _____）という言葉がある。膿胸は別名（④ _____）とも言う。

   a) empyema   b) thorax   c) pyothorax   d) pus

2. 馬の喘鳴症（① _____）、または喉頭（片）麻痺（② _____）は喉頭部の機能障害（③ _____）により、喉頭（④ _____）の部分的閉塞をきたす疾患である。
   主に3～6歳の競走馬におこるが、中～老年の大型犬に発生することもある。反回神経（⑤ _____）が麻痺し、その支配下にある喉頭筋の変性萎縮がおこることによる。馬では、90％以上が左側の片側麻痺である。心臓および大動脈（⑥ _____）起始部の過機能性肥大によるとされている。

   a) recurrent laryngeal nerve   b) dysfunction
   c) laryngeal paralysis（laryngohemiplegia）
   d) roaring   e) larynx   f) aorta

Exercises 6 の答え

問1．① rhinitis ② laryngitis ③ tracheitis ④ pleuritis ⑤ pneumonia（pneumonitis）
問2．① apnea ② dyspnea ③ eupnea ④ hypopnea ⑤ hyperpnea
問3．1. ① c) ② b) ③ d) ④ a) 2. ① d) ② c) ③ b) ④ e) ⑤ a) ⑥ f)

Chapter 8

消化器系とその疾患
The Digestive System and its Disorders

1 消化器系 The Digestive System

a) 消化と消化管 Digestion and the Alimentary Canal

alimentary canal（消化管）の alimentary はラテン語の alimentum（食物）に由来する。消化管は digestive tract とも呼ばれる。gastrointestinal tract（胃腸管）は、しばしば GI tract と略される。

消化管　Alimentary Canal 🔊

aliment, nourishment	栄養物、食物
alimentary	栄養の
alimentary canal	消化管
alimentation	栄養法（栄養を与えること）

消化　Digestion 🔊

in + gest*¹	ingest／ingestion	食物などを摂取する／通常飲み込んで、食物や薬剤を体内に摂取すること
di + gest*²	digest／digestion／digestive	消化する／消化／消化を助ける
in + digest*³	indigestion／indigested／indigestibility／indigestible	不消化、消化不良／不消化の／不消化／消化しにくい

*1　ラテン語の in-（内へ）＋ gest-, gerere（運ぶ）。
*2　ラテン語の di-（切り離す）＋ gest-, gerere（運ぶ）。
*3　in-（否定）＋ digest。

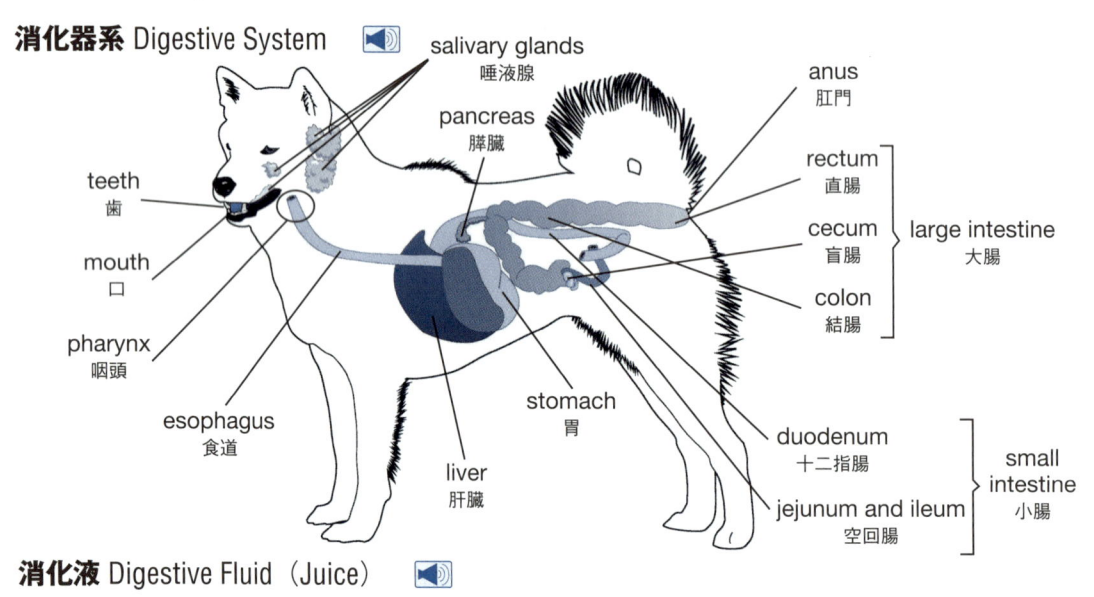

消化器系 Digestive System 🔊

（図：犬の消化器系）
- salivary glands 唾液腺
- pancreas 膵臓
- anus 肛門
- rectum 直腸
- cecum 盲腸
- colon 結腸
- large intestine 大腸
- teeth 歯
- mouth 口
- pharynx 咽頭
- esophagus 食道
- liver 肝臓
- stomach 胃
- duodenum 十二指腸
- jejunum and ileum 空回腸
- small intestine 小腸

消化液 Digestive Fluid（Juice）🔊

saliva	唾液	pH 6.7
bile	胆汁	pH 7.1-8.5
gastric fluid, gastric juice *	胃液	pH 2.0
pancreatic fluid, pancreatic juice *	膵液	pH 8.4
intestinal fluid, intestinal juice *	腸液	pH 8.0

*　-juice を含む消化液名は一般用語。

b) 口腔 Oral Cavity
口 Mouth

語幹	英語名詞	英語形容詞	ラテン語名詞	ギリシャ語名詞
oral-, stom-, stomat(o)-	mouth	oral, stomatic	os(sg.), ora(pl.), oris	stoma(sg.), stomata(pl.)

　ラテン語の os には「口」と「骨」というふたつの意味がある。per os（経口投与）という時の os は骨ではない。
　ギリシャ語の stoma はこのまま英語として、「小孔」や「瘻」を意味する医学用語でもある。

口にかかわる用語 Oral Terminology 🔊

mouth	foot-and-mouth-disease	口蹄疫
oral-	**oral**e（=oral point）	オラーレ[*1]
stom-, stomat(o)-	**stomat**algia, **stomat**odynia	口腔痛
	stomatitis（=canker sores）	口内炎
	di**stoma**	ジストマ[*2]
その他	glossitis	舌炎
	tonsillitis	扁桃炎
	parotitis	耳下腺炎

*1　oral point（切歯縫合の歯槽端舌側面の点）とも言う。
*2　吸虫、二口虫ともいう。di（2）＋stoma（口）。

食欲の異常および渇欲の異常 Abnormal Appetite and Morbid Thirst 🔊

dipsesis	高度口渇*
polydipsia	渇欲亢進、多渇症
disturbance of mastication and deglutition	咀嚼および嚥下障害
disturbance of rumination	反すうの障害
bloat, disturbance of eructation relex	鼓脹
vomiting	嘔吐

*　ギリシャ語の dipseo（のどが渇くこと）より。

唾液 Saliva

語幹	英語名詞	英語形容詞	ラテン語名詞	ギリシャ語名詞
sial(o)-, ptyal(o)-	saliva	salivary	saliva	sialon, ptyalon

唾液にかかわる用語 Salivary Terminology 🔊

saliva	saliva／salivate／salivary／salivation	唾液／唾液を出す／唾液の／唾液の分泌
sial(o)-, ptyal(o)-	**sial**ism, **sial**ismus, **sialo**rrhea, **ptyal**ism	流涎症
	sialogram, **ptyalo**graphy	唾液腺造影
	sialolith, **ptyalo**lith	唾石
	sialodochitis*	唾液管炎

*　doch(o)- は主管を意味し、choledoch（総胆管）の doch と同じ。

c) 歯 Teeth 🔊

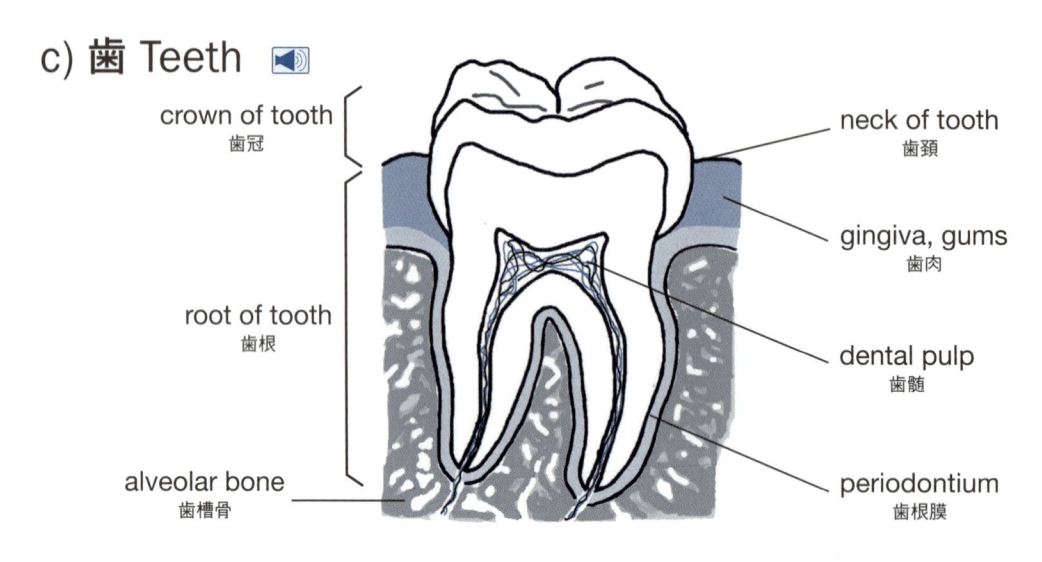

- crown of tooth 歯冠
- root of tooth 歯根
- alveolar bone 歯槽骨
- neck of tooth 歯頸
- gingiva, gums 歯肉
- dental pulp 歯髄
- periodontium 歯根膜

語幹	英語名詞	英語形容詞	ラテン語名詞	ギリシャ語名詞
odont(o)-	tooth(sg.), teeth(pl.)	dental	dens(sg.), dentes(pl.)	odous

歯にかかわる用語 Dental Terminology 🔊

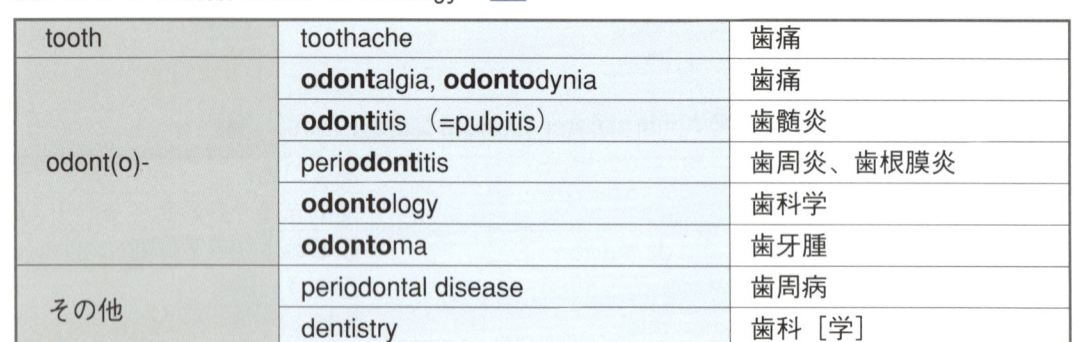

tooth	toothache	歯痛
	odontalgia, **odonto**dynia	歯痛
odont(o)-	**odont**itis （=pulpitis）	歯髄炎
	peri**odont**itis	歯周炎、歯根膜炎
	odontology	歯科学
	odontoma	歯牙腫
その他	periodontal disease	歯周病
	dentistry	歯科［学］

咀嚼 Mastication 🔊

masticate	噛む、咀嚼する
mastication	咀嚼

歯の疾患 Teeth Disorders 🔊

decayed teeth, carious teeth, cavity, dental caries *1	虫歯
fill a cavity	虫歯に詰めものをする
pull out a tooth	抜歯
an artificial tooth, a set of false teeth, dentures *2	入れ歯
a tooth mark	（噛み付かれた時の）歯型

*1 用法例：When we have decayed teeth or carious teeth, we go see a dentist to fill the cavity.（虫歯になると歯医者へ行き、詰めものをしてもらう。）

*2 用法例：A set of artificial teeth or false teeth is also called dentures, although the former is much common.（「人工的な歯」あるいは「偽歯」は「入れ歯」とも呼ばれる。前者の方が一般的に用いられる。）

エナメル質とセメント質 Enamel and Cement 🔊

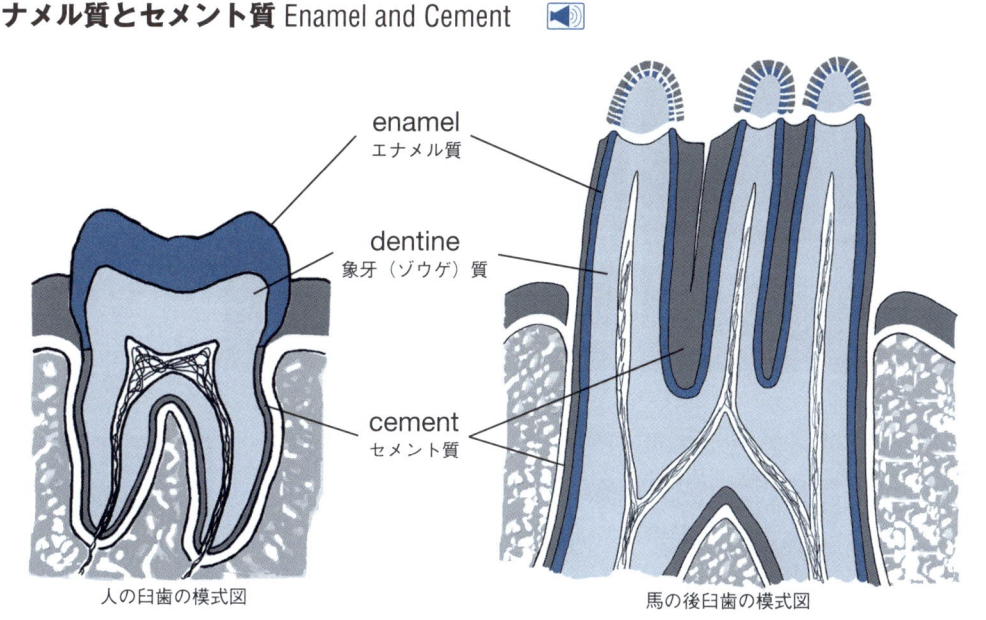

- enamel エナメル質
- dentine 象牙（ゾウゲ）質
- cement セメント質

人の臼歯の模式図　　　　馬の後臼歯の模式図

　馬の歯はとても複雑な形をしているように見えるが、エナメル質、ゾウゲ質、およびセメント質がヒダを作って、その上部（破線の部分）が摩滅してなくなった、と考えれば理解しやすい。
　摩滅する時、比較的軟らかいゾウゲ質やセメント質は早く磨り減り、一番硬いエナメル質が残る。その結果、咬合面に複雑なエナメル稜（enamel crest）が露出する。

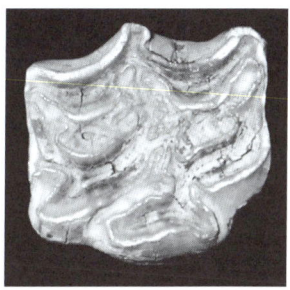

馬の後臼歯

歯肉 Gingiva

語幹	英語名詞	英語形容詞	ラテン語名詞	ギリシャ語名詞
gingiv(o)-	gingiva(sg.), gingivae(pl.)	gingival	gingiva	-

　歯肉の医学用語は gingiva だが、一般用語は gums である。単数形は gum だが通常は複数形を用いる。

歯肉と歯槽の疾患　Gingiva, Alveolus and their Disorders 🔊

gingiva 歯肉	gingivitis [*1]	歯肉炎
	epulis	歯肉腫、エプーリス
alveolus 歯槽	alveolitis [*2]	歯槽炎
	alveolar osteitis	歯槽骨炎
	alveolar pyorrhea [*3]	歯槽膿漏

[*1]　gingivitis = gingiv(o)- + -itis。
[*2]　歯を入れる歯槽骨は alveolar bone。この alveolar には「歯槽の」という意味の他に「肺胞の」という意味もある。歯槽の一般用語は tooth socket。
[*3]　py(o)-（膿）+ -rrhea（漏出）より。

猿の歯 Teeth of Monkey 🔊

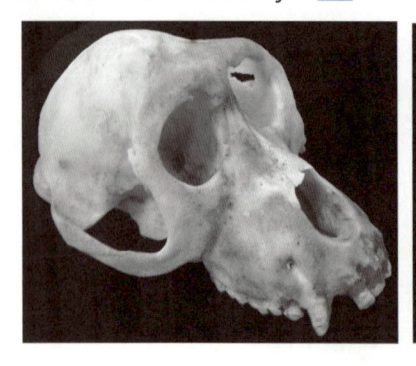

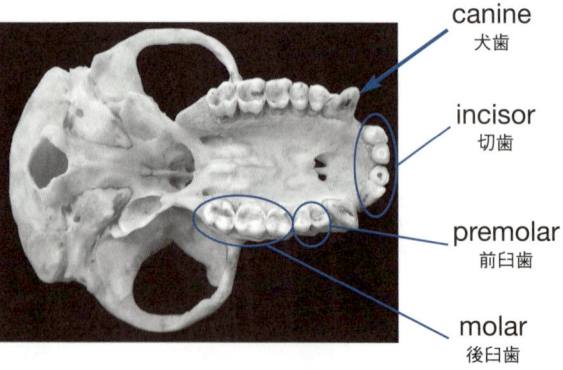

canine 犬歯
incisor 切歯
premolar 前臼歯
molar 後臼歯

イノシシの牙 Tusks of Wild Boar 🔊

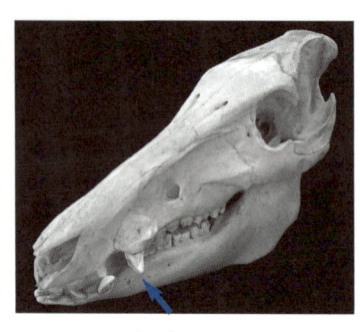

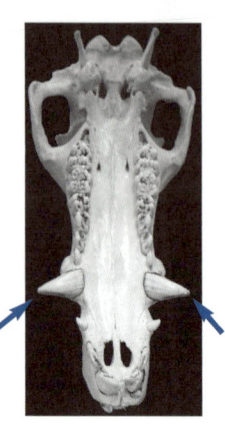

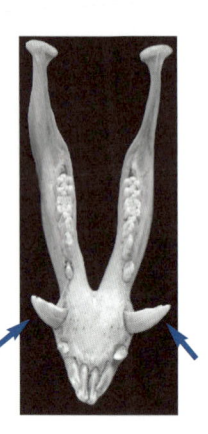

tusk 牙

人と動物の歯科用語の比較 Comparison of Human and Animal Dental Terms 🔊

	医学	獣医学
incisor	門歯	切歯
canine	犬歯	犬歯
premolar	小臼歯	前臼歯
molar	大臼歯	後臼歯
deciduous teeth, milk teeth, temporary teeth	乳歯	乳歯
permanent teeth	永久歯	永久歯

牙 Tusk and Fang 🔊

tooth(sg.), teeth(pl.)	歯
tusk (sg.), tusks(pl.)	牙*1
fang(sg.), fangs(pl.)	牙*2

*1 象やイノシシの牙のように、口の外にはみ出している歯。必ずしも犬歯由来とは限らない。例えば象の牙は切歯である。
*2 食肉類の犬歯。スラングなので、獣医師は canine teeth を使う。

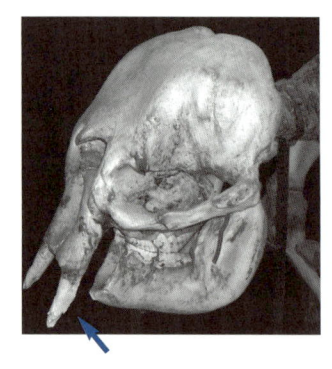

象の子供のtusk（大人になるともっと伸びる）。

d) 咽頭、食道　Pharynx and Esophagus

咽頭 Pharynx*1

語幹	英語名詞	英語形容詞	ラテン語名詞	ギリシャ語名詞
pharyng(o)-	pharynx	pharyngeal	pharynx	pharynx*2

*1　口腔と咽頭の間の軟口蓋と舌根に囲まれた空間を指す、fauces（口峡）という類語あり。ラテン語の「のど」より。
*2　「のど」という意味。

食道 Esophagus

語幹	英語名詞	英語形容詞	ラテン語名詞	ギリシャ語名詞
esophag(o)-	esophagus (sg.), esophagi (pl.)	esophageal	oesophagus, esophagus	oisophagos*

*　oisein（運ぶ）＋ phagos（食べる）より。ちなみに macrophage（大食細胞）の phage も phagos に由来。

咽頭および食道にかかわる用語 Terminology of the Pharynx and Esophagus

語幹	用語	和訳
pharyng(o)-	**pharyng**itis （=angina）	咽頭炎（アンギナ）
pharyngeal	pharyngeal paralysis	咽頭麻痺
esophag(o)-	**esophag**itis	食道炎
	reflux **esophag**itis *	逆流性食道炎
esophagus	dilatation of esophagus （=**esophag**ectasia, **esophag**ectasis）	食道拡張
	megaesophagus	巨大食道［症］
	paralysis of esophagus （=achalasia）	食道麻痺（アカラシア）
	spasm of esophagus	食道痙攣
esophageal	esophageal adenocarcinoma	食道腺癌
	esophageal diverticulum	食道憩室
	esophageal stenosis	食道狭窄
	esophageal obstruction	食道閉塞

*　Barrett's esophagus（バレット食道）はこの合併症である。

> **Column　口から胃へと変化する言葉**
>
> 　　古代ギリシャ語の stoma は口を意味した。これが転じて stomakhos（のど、食道）となり、ラテン語の stomachus（のど、食道、胃）となり、フランス語古語 stomaque をへて、英語の stomach になった。ひとつの言葉の意味が、口からのど、食道、胃へとだんだんとずり下がってくるところが面白い。

e) 腹腔と骨盤腔 Abdominal and Pelvic Cavities

腹腔 Abdominal Cavity 🔊

「内臓」「腹」など関連用語は多い。一般用語に gut がある。これは胃腸全体、腹腔臓器全体を、漠然と指す言葉である。

語幹	英語名詞	英語形容詞	ラテン語名詞	ギリシャ語名詞	和訳
splanchn(o)-, splanchni-	-	splanchnic	-	splanchnon	内臓
viscer(o)-	viscus	visceral	viscus(sg.), viscera(pl.)[*1]	-	内臓
celi(o)-	-	celiac	-	coilos[*1]	腹、腹腔、腸
abdomin(o)-	abdomen	abdominal	abdomen	-	腹腔
laparo-	-	-	-	lapara[*1]	一般に腹
peritone(o)-	peritoneum	peritoneal	peritoneum	peritonation[*2]	腹膜

*1 viscera（体の内部）、coilos（空洞）、lapara（わき腹）を意味する。
*2 peri（周囲）+ tonos（伸展した）より。

腹腔にかかわる用語 Terminology of the Abdominal Cavity 🔊

語幹		和訳
splanchn(o)-	**splanchno**logy	内臓学
viscer(o)-	**viscero**motor neuron	内臓運動ニューロン
	viscerosensory	内臓感覚の
celi(o)-	**celio**centesis	腹腔穿刺
	celiorrhaphy	腹壁縫合
celiac	celiac artery	腹腔動脈
abdomin(o)-	**abdomino**scopy（=**peritoneo**scopy）	腹腔鏡検査［法］
abdominal	abdominal cavity	腹腔
	abdominal wall	腹壁
	abdominal fluid（=**hydroperitone**um, ascites*）	腹水
laparo-	**laparo**tomy（=celiotomy）	開腹術
	laparocele（=abdominal hernia）	腹部ヘルニア
peritone(o)-	**periton**itis	腹膜炎
	feline infectious **periton**itis（FIP）	猫伝染性腹膜炎
	diffuse **periton**itis	びまん性腹膜炎
	peritoneopericardial diaphragmatic hernia（PPDH）	腹膜心膜横隔膜ヘルニア

* askos + ites（〜様の）より。askos とはラテン語で革製の酒袋をいう。腹水で膨らんだ状態が似ていることから名付けられた。

骨盤 Pelvis

語幹	英語名詞	英語形容詞	ラテン語名詞	ギリシャ語名詞
pervi(o)-, pelvo-	pelvis*	pelvic	pelvis	-

* pelvis（骨盤）の複数は pelves。ラテン語の水盤、鉢、洗面器のような浅い盤という意味。

f) 胃 Stomach 🔊

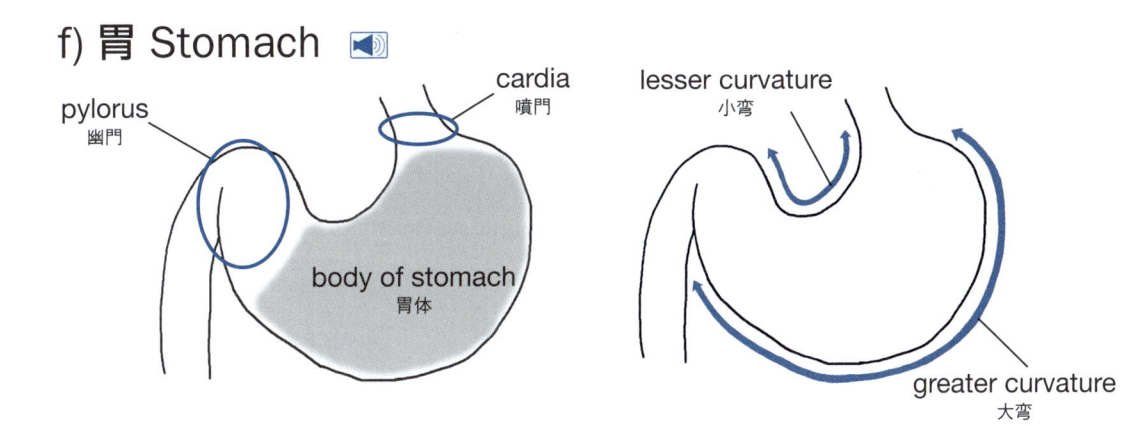

語幹	英語名詞	英語形容詞	ラテン語名詞	ギリシャ語名詞
gastr(o)-	stomach	gastric	ventricle	gastēr

　ラテン語の ventricle は、袋状のものを指す。心臓の心室も ventricle であり、脳の脳室も ventricle である。まぎらわしいので注意しよう。

胃にかかわる用語 Stomach Terminology 🔊

stomach	stomachache	胃痛
	stomachic	胃の、胃に良い、健胃剤
	stomach tube	胃管、胃ゾンデ
	stomach worm	捻転胃虫
	stomach pump	胃洗浄器、胃ポンプ
	stomach cancer（=gastric cancer）	胃癌
	stomach ulcer（=gastric ulcer）	胃潰瘍
	stomach flu（=gastric flu[*1], **gastro**enteritis）	胃腸の炎症
gastr(o)-	**gastr**algia, **gastro**dynia	胃痛
	gastritis	胃炎
	gastrectomy	胃切除術
	chronic atrophic **gastr**itis	慢性萎縮性胃炎
	gastrocamera	胃カメラ[*2]
gastric	gastric irrigation, gastric lavage	胃洗浄
	gastric dilation volvulus（GDV）	胃軸捻転
	gastric torsion	胃捻転
	gastric dilatation	胃拡張

[*1]　インフルエンザとは関係がないのに、しばしば gastric flu あるいは stomach flu とも呼ばれる。
[*2]　内視鏡は endoscopy。

胃腺 Gastric Glands 🔊

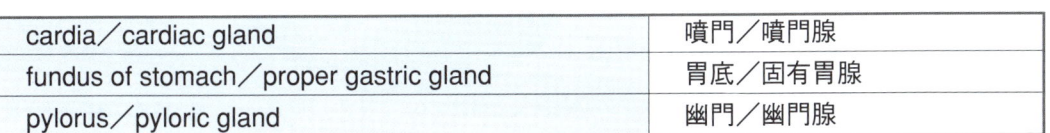

cardia／cardiac gland	噴門／噴門腺
fundus of stomach／proper gastric gland	胃底／固有胃腺
pylorus／pyloric gland	幽門／幽門腺

噴門と幽門 Cardia and Pylorus

　胃と心臓のもうひとつの共有単語は、cardia（噴門）である。これはもともと、噴門が、心臓の近くに位置することから付けられたとされる。cardiac muscle というとふつう、心筋を指すが、噴門筋も同じく cardiac muscle なのである。

cardia 噴門	epicardia	噴門上部
	cardio chalasia	噴門痙攣
pylorus 幽門	pyloric dysfunction （=pyloric insuffiency）	幽門機能不全
	pyloric stenosis	幽門狭窄
	pylorospasm	幽門痙攣

胃液 Gastric Fluid

acid	stomach acid	胃酸
	hydrochloric acid	塩酸（HCl）、胃酸の成分
	hyperacid	胃酸過多の
	hyperacidity	胃酸過多症
pep-*1	pepsin	ペプシン
	peptic	ペプシンの、消化力を有する
	dyspepsia	消化不良*2
	dyspeptic	消化不良の
	peptic ulcer（=peptic ulcer disease［PUD］）	胃、十二指腸の消化性潰瘍

*1　ギリシャ語の pepsis（消化）より。
*2　胃腸全体を指す。胃だけの症状ではない。

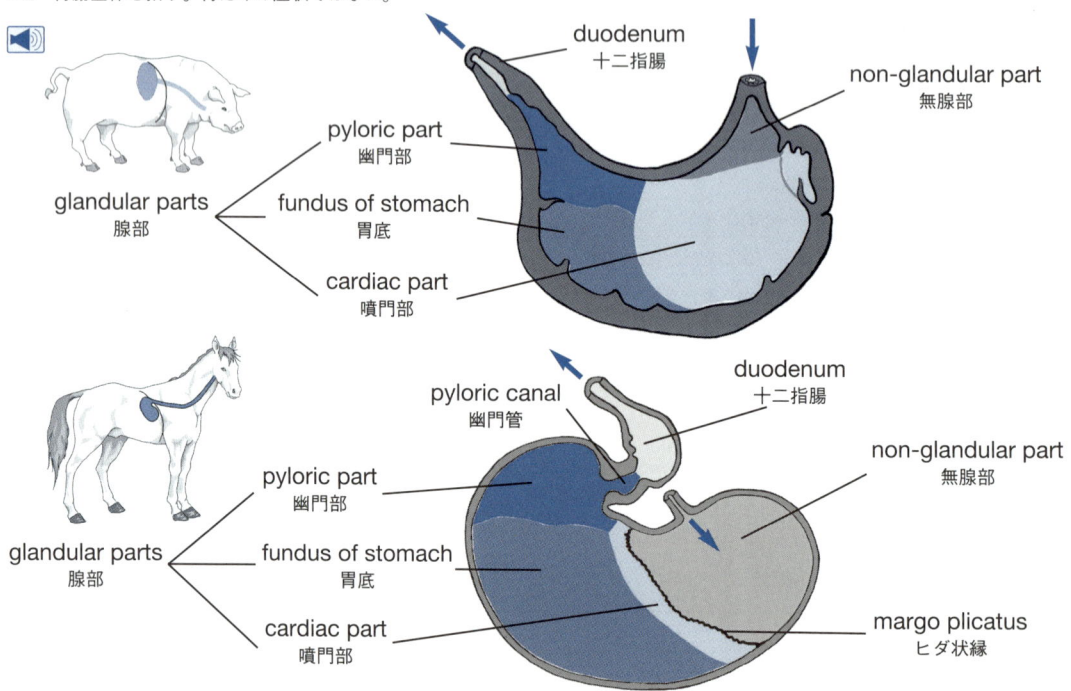

人や犬の胃は無腺部が無い。豚や馬も外見は単胃だが、解剖すると噴門近くに、反すう類の第一〜三胃と同様、重層扁平上皮を持つ無腺部がある。すなわち、複胃との移行形である。

g) 反すう胃 Ruminant Stomachs

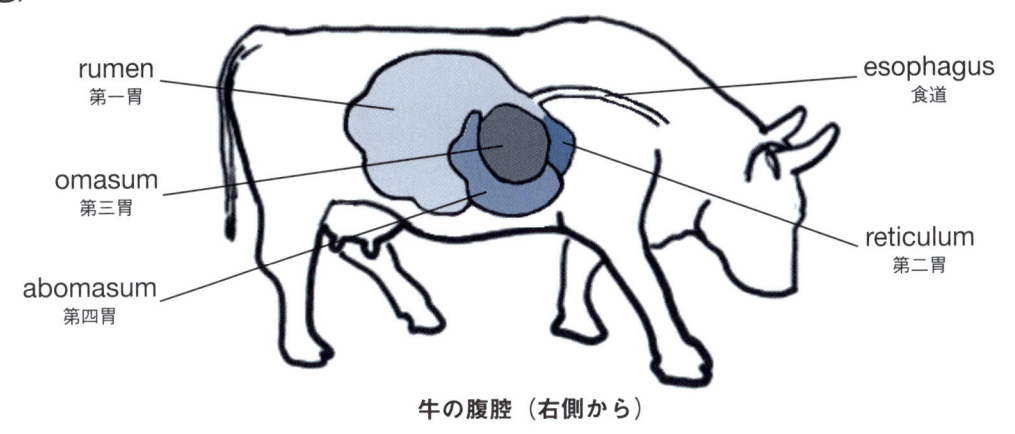

牛の腹腔（右側から）

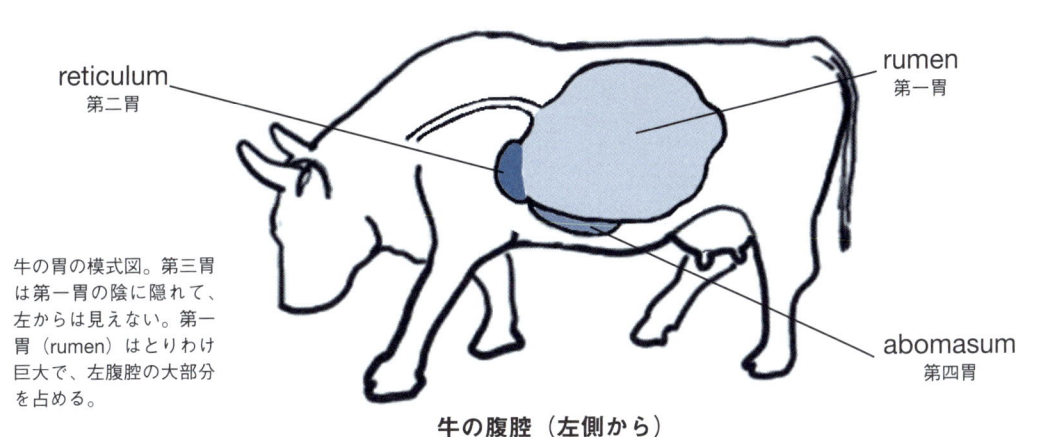

牛の腹腔（左側から）

牛の胃の模式図。第三胃は第一胃の陰に隠れて、左からは見えない。第一胃（rumen）はとりわけ巨大で、左腹腔の大部分を占める。

第一胃 Rumen 🔊

語幹	英語名詞	英語形容詞	ラテン語名詞	ギリシャ語名詞
rumin(o)-, rumen(o)-	rumen (sg.), rumina, rumens (pl.)	ruminal	rumen	-

「反すうする」という英語動詞の ruminate はラテン語の ruminatus（ruminare〔反すうする〕の過去分詞）に由来する。英語の rumen は、ラテン語の同じく rumen に由来するが、ラテン語の本来の意味はのどや食道だった。それが胃の方へと次第にずれていき、現在の反すう類の第一胃を意味するようになった。

反すう胃 Ruminant Stomachs 🔊

ruminate	反すうする
rumination	反すう
ruminant	反すう動物
rumen	第一胃
reticulum	第二胃
omasum	第三胃
abomasum	第四胃

反すう胃の疾患 Diseases of Ruminant Stomachs

atony of the forestomach		前胃弛緩症
vagal indigestion		迷走神経性消化不良
rumen 第一胃	ruminitis	第一胃炎
	acute ruminal dilatation	第一胃急性拡張
	ruminal impaction	第一胃食滞
	ruminal tympany, bloat	第一胃鼓脹症
	rumen acidosis	第一胃過酸症 （ルーメン・アシドーシス）
	rumen alkalosis	ルーメン・アルカローシス
	rumen putrefaction	第一胃腐敗症
	ruminal parakeratosis	ルーメン・パラケラトーシス
	rumenotomy	第一胃切開術
reticulum 第二胃	reticulitis	第二胃炎
	traumatic reticulitis	創傷性第二胃炎
	traumatic reticuloperitonitis	外傷性（創傷性）第二胃腹膜炎
omasum 第三胃	omasitis	第三胃炎
	obstruction of omasum	第三胃梗塞
	omasal impaction	第三胃食滞
abomasum 第四胃	abomasitis	第四胃炎
	eroison and ulcer of abomasum	第四胃のびらんおよび潰瘍
	atony or impaction of abomasum	第四胃弛緩症または第四胃食滞
	displaced abomasum	第四胃変位
	torsion of abomasum	第四胃捻転

h) 腸管 Intestinal Tract 🔊

右図：犬の消化器系。空回腸は犬種にもよるが、2.0～4.5メートルもあり、左腹腔の大部分を占める。
空回腸を手で持ち上げることにより、これに隠されている右側の盲腸、結腸、直腸をのぞき見ることができる。

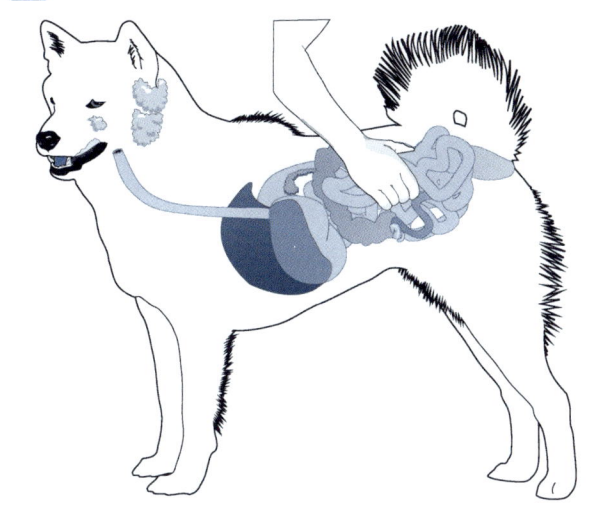

右図：空回腸の一部を切断した模式図。切断により、結腸の全走行、十二指腸と結腸の位置関係、盲腸、直腸がよく見えるようになる。

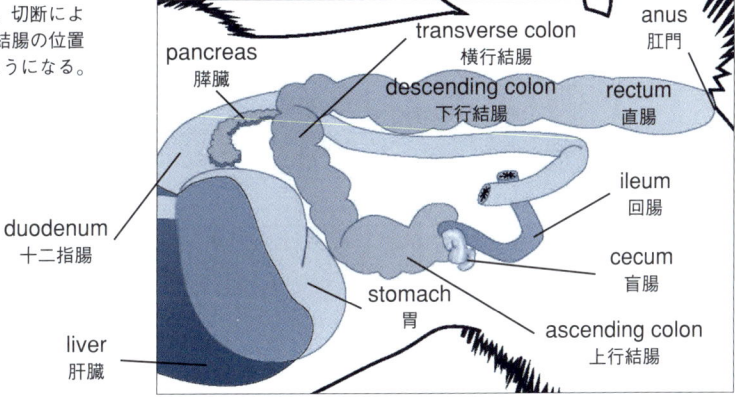

腸 Intestine

語幹	英語名詞	英語形容詞	ラテン語名詞	ギリシャ語名詞
intestin-, enter(o)-	intestine	intestinal	intestinum[*1]	enteron
-	bowel	-	botellus[*2]	-

腸の英語は intestine。英語の一般会話では intestinal tract、と2語で言い表すことが多い。他に entrails（はらわた、臓物）という言葉もある。しばしば腸と同義語で用いられる。この entrails（常に複数形）は、ラテン語の interanea（interaneum、腸の複数形）から派生した。

[*1] ラテン語の intestinum は intestinus の中性形。intestinus は intus（～の中）より派生した。
[*2] bottelus はもともと「小さなソーセージ」の意味。

腸にかかわる用語 Intestinal Terminology

intestnal	intestinal	腸の
	intestinal flora	腸内細菌叢
	intestinal fluid	腸液
	intestinal anastomosis	腸吻合術
	gastrointestinal tract（GI tract）	胃腸管、消化管の胃腸の部分
intestine	large intestine（=large bowel）	大腸
	small intestine（=small bowel）	小腸
bowel	bowel sounds	腸音*1
	inflammatory bowel disease（IBD）	炎症性腸疾患
	bowel movement（BM）	排便
enter(o)-	mes**entery**	腸間膜*2
	enteralgia, **enter**dynia, **entero**dynia	腸痛
	enteritis	腸炎
	swine transmissible gastro**enter**itis	豚伝染性胃腸炎
	dys**entery**	赤痢
その他	volvulus, twisting of the intestines	腸捻転
	intussusception	重積*3

*1 　腸内を内容物が送られることにより生じる、比較的高い音。
*2 　mes はギリシャ語の mesos（中間）より。
*3 　intus（内）+ suscipio（取る）より。

i) 小腸 Small Intestine

十二指腸 Duodenum

語幹	英語名詞	英語形容詞	ラテン語名詞	ギリシャ語名詞
duoden(o)-	duodenum (sg.), duodena (pl.)	duodenal	duodenum	-

　duodenum の duo はラテン語の duodeni（12）に由来する。指の腹の幅12個分、の意味である。「12本（ラテン語で duodenum ）の指の幅の腸」と記載したのは16世紀の解剖学者 Vesalius で、その著書「人体構造論」の中でのことである。人では約25cmだが、馬や牛では1m以上もある。

空腸 Jejunum

語幹	英語名詞	英語形容詞	ラテン語名詞	ギリシャ語名詞
jejun(o)-	jejunum (sg.), jejunums (pl.)	jejunal	jejunum	-

　jejunum はラテン語の jejunus（空の、空腹の）に由来する。死後遺体を解剖すると、たいてい空だったためである。死んだ直後も蠕動（peristalsis）がしばらく続いて内容物を回腸の方に押し出した後だったか、あるいはそもそも死ぬ直前には食欲も失せ、ものを食べないことが多いので空だったと考えられる。

　もともとは紀元前2世紀、古代ギリシャの偉大な医学者ガレノス（Galenus, 英語は Galen）が、この臓器をギリシャ語の nestis（断食）と名付けたことに始まる。これがラテン語に取り入れられ、似た意味の単語を当てて jejunum と訳された。

　ラテン語の jejunus からは、別の英語 jejune も派生した。jejune は未熟な、子供じみた、幼稚な、不毛な、無味乾燥な、という形容詞である。

回腸 Ileum

語幹	英語名詞	英語形容詞	ラテン語名詞	ギリシャ語名詞
ileo-	ileum (sg.), ileums (pl.)	ileal, ileac	ileum	-

　ileum は、腸骨 ilium（複数形 ilia ）と同様、ラテン語の ilia（鼡径部、わき腹）に由来する。人では空腸（約2.5m）より回腸の方が長い（2〜4m）が、動物では回腸は比較的短く、空腸の方が回腸よりはるかに長い。草食獣の空回腸の総長は極めて長く、牛では40mにも達する。

小腸にかかわる用語 Terminology of the Small Intestine*

duoden(o)-	**duoden**itis	十二指腸炎
jejun(o)-	**jejun**itis	空腸炎
ileo-	**ile**itis	回腸炎

* ileum（回腸）、ilium （腸骨）、ileus（イレウス、腸閉塞）、互いにスペルが良く似た単語である。イレウスはギリシャ語の eileos より。eileos は同じくギリシャ語の eilein（絞る、きつく巻く）の派生語。

イレウス（腸閉塞） Ileus

Ileus	paralytic ileus, adynamic ileus	麻痺性腸閉塞、麻痺性イレウス
	mechanical ileus	機械性腸閉塞、機械性イレウス
	adhesive ileus	癒着性腸閉塞

j) 大腸 Large Intestine

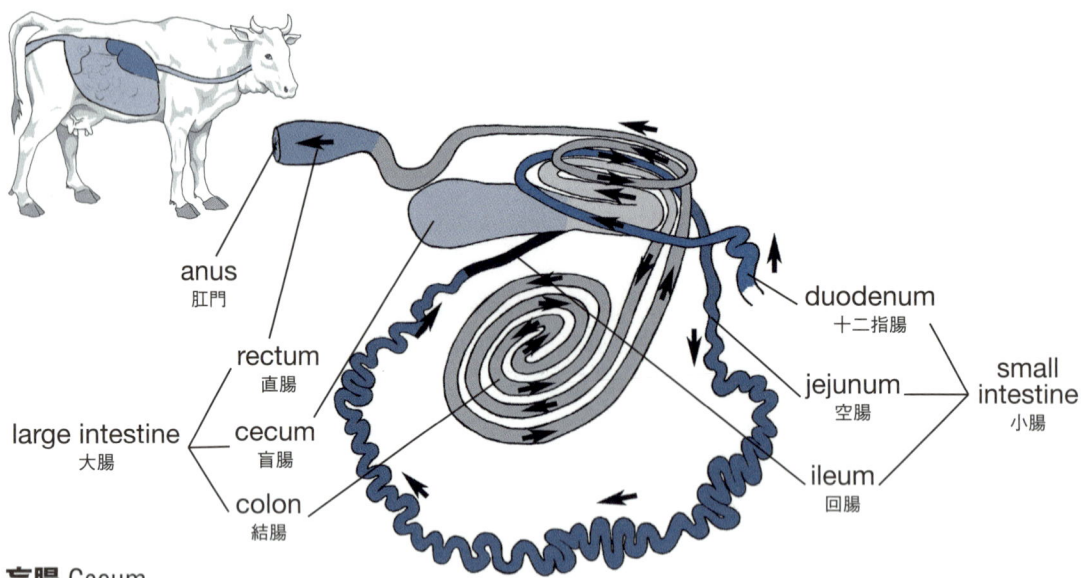

盲腸 Cecum

語幹	英語名詞	英語形容詞	ラテン語名詞	ギリシャ語名詞
caec(o)-, cec(o)-	cecum (sg.) caeca, ceca (pl.)	caecal, cecal	caecum, cecum	-

　cecum（盲腸）はラテン語の intestinum caecum より。caecum は blind（盲）という意味の caecus が語尾変化したもの。この腸は盲端に終わることからこの名が付いた。

虫垂（盲腸の先端部） Appendix

語幹	英語名詞	英語形容詞	ラテン語名詞	ギリシャ語名詞
appendic(o)-	appendix (sg.) appendices (pl.)	appendical, appendiceal	appendix	-

　appendix（虫垂）は、ラテン語の appendage より。この言葉は、ad-（〜へ）+ pendere（吊るす、掛ける）より成る appendere の派生語である。もともとの言葉の意味には「垂」はあるが、「虫」はない。

結腸 Colon

語幹	英語名詞	英語形容詞	ラテン語名詞	ギリシャ語名詞
col(o)-	colon (sg.) colons, cola (pl.)	colic, colonic	colon	kolon

直腸 Rectum

語幹	英語名詞	英語形容詞	ラテン語名詞	ギリシャ語名詞
rect(o)-	rectum (sg.) rectums, recta (pl.)	rectal	rectum	-

　rectum はラテン語の rectum（ストレート）に由来する。

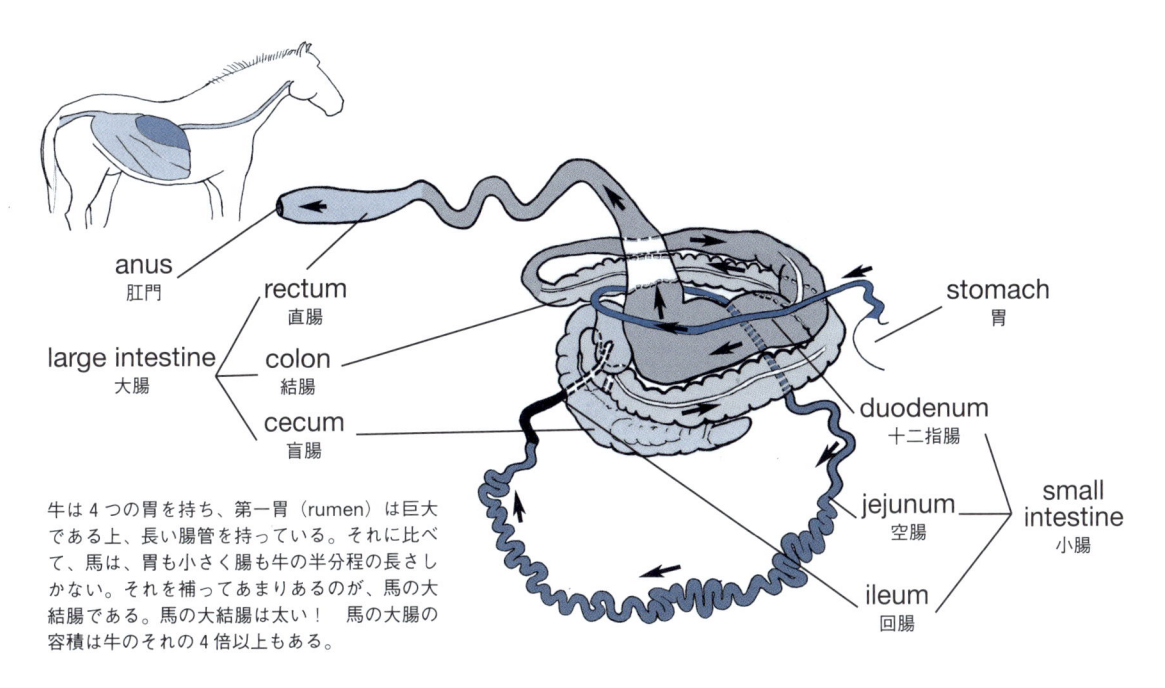

牛は4つの胃を持ち、第一胃（rumen）は巨大である上、長い腸管を持っている。それに比べて、馬は、胃も小さく腸も牛の半分程の長さしかない。それを補ってあまりあるのが、馬の大結腸である。馬の大結腸は太い！ 馬の大腸の容積は牛のそれの4倍以上もある。

大腸にかかわる用語 Terminology of the Large Intestine

cec(o)-	**cec**itis （=typhlitis）	盲腸炎
appendic(o)-	**appendic**itis	虫垂炎
	appendectomy	虫垂切除術
col(o)-	**col**itis*1	結腸炎
	coliemia	大腸菌血症
	Escherichia coli（E. coli）	大腸菌
	colospasm	結腸痙攣
colic, colonic	colonic diverticula （=diverticula of colon）	結腸憩室
	colic*2	疝痛
colon	megacolon	巨大結腸症
rect(o)-	**rect**itis （=proctitis）	直腸炎
rectal	rectal cancer	直腸癌
	rectal prolapse	直腸脱
rectum	polyps of rectum	直腸ポリープ

*1 colitis という言葉は、結腸炎だけに限らずもっと広範囲に「大腸炎」という意味でも使われる。
*2 colic は colon（結腸）の形容詞であると同時に、「疝痛」を意味する。疝痛の colic は、ラテン語の colicus（結腸の病気）、ギリシャ語の kolikos（kolon の派生語）に由来する。ギリシャ語でもラテン語でも、colic を「結腸の病気」と考えていたことが分かる。

k) 肛門 Anus

語幹	英語名詞	英語形容詞	ラテン語名詞	ギリシャ語名詞
proct(o)-	anus(sg.), anuses, ani(pl.)	anal	anus	proctos

肛門にかかわる用語 Anal Terminology 🔊

proct(o)-	**proct**ocele（=**rect**ocele）	直腸脱
	proctectomy	直腸切除術
	proctostomy*¹	人工肛門形成術
anal	anal fistula	肛門裂傷、裂肛、肛門瘻、痔瘻
	perianal gland tumor（=hepatoid gland carcinoma）	肛門周囲腺腫瘍*² （肝様腺腫、肝様腺癌）
	perianal fistula	肛門周囲瘻孔
その他	hemorrhoid	痔

*1 stoma はギリシャ語の「口」。
*2 肛門周囲を構成する細胞は大きさ、形、染色性が肝細胞に似ていることから肝様腺とも言われる。すなわち肛門周囲腺腫瘍と肝様腺腫はほとんど同義語である。

排便 Defecating 🔊

diarrhea	下痢
constipation	便秘
suppository	坐薬
enema	浣腸*

* ギリシャ語の enema（注入）。

2 肝臓と胆嚢 The Liver and Gall Bladder

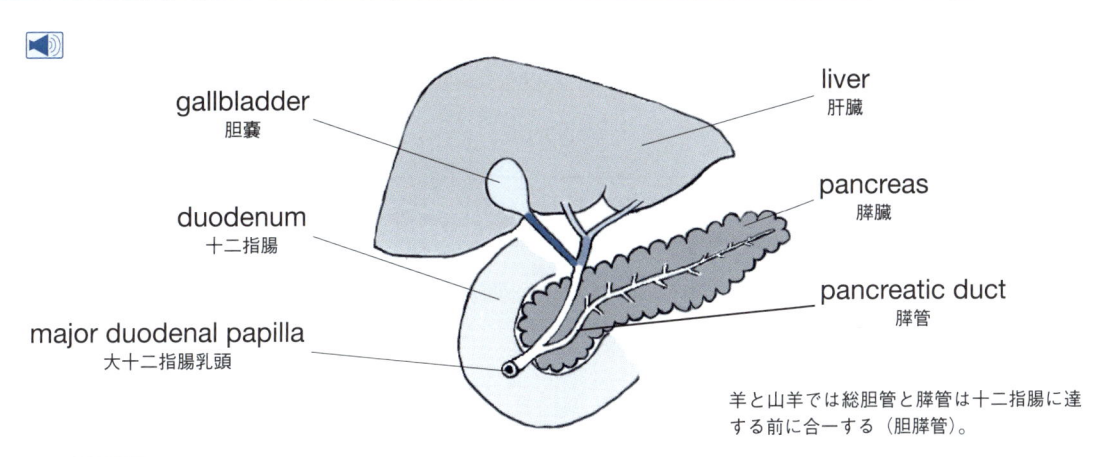

羊と山羊では総胆管と膵管は十二指腸に達する前に合一する（胆膵管）。

a) 肝臓 Liver

語幹	英語名詞	英語形容詞	ラテン語名詞	ギリシャ語名詞
hepat(o)-	liver	hepatic	hepar	hēpar, hēpatos*

* ラテン語の属格（＝英語の所有格）。

肝臓にかかわる用語 Hepatic Terminology

liver	liver cancer, tumors of the liver	肝臓癌、肝腫瘍
	fatty liver （=hepatic steatosis）	脂肪肝
	nutmeg liver	ニクズク肝（慢性うっ血肝）
	congestion of the liver	うっ血肝
hepatic	hepatic duct	肝管
	common hepatic duct	総肝管
	hepatic lobule	肝小葉
	hepatic failure	肝不全
	hepatic coma	肝性昏睡
hepat(o)-	**hepato**cyte	肝細胞
	hepatophyma	肝腫瘤*1
	hepatitis 肝炎 ／ viral **hepat**itis	ウイルス性肝炎
	hepatitis 肝炎 ／ non-alcoholic steato**hepat**itis（NASH）	非アルコール性脂肪性肝炎
	hepatitis 肝炎 ／ infectious canine **hepat**itis	イヌ伝染性肝炎
	hepatomegaly	肝腫大
	Distoma **hepat**icum（liver-fluke）	肝蛭（肝吸虫）
その他	cirrhosis*2	肝硬変
	alcoholic cirrhosis	アルコール性肝硬変
	heparin	ヘパリン
	jaundice, icterus*3	黄疸

*1 肝臓の球形または結節性腫瘤。
*2 肝硬変（cirrhosis）はギリシャ語の kirrhos（黄褐色）にちなんで、フランス人医師 Laennec（1781-1826）により命名された。Laennec は、メラノーマ（黒色腫）の名付けの親であり、stethoscope（聴診器）を発明した人物でもある。
*3 jaune はフランス語の「黄色」、icterus はギリシャ語の ikteros（胆汁）より。

胆汁 Bile

語幹	英語名詞	英語形容詞	ラテン語名詞	ギリシャ語名詞
chol-, chole-, cholo-, bili-	bile, gall	biliary	bilis	cholē

　gall（胆汁）は中世英語の galle に由来する。bile（胆汁）と同じ意味であるが、gall の方が古めかしい表現である。獣医学分野では動物に対してまだ使われることもある。胆汁がたいへん苦い物質であることから、gall は、苦々しい思い、憎しみ、厚かましさなどをも意味する。なお胆汁を表す英語は、人体医学分野では主として bile が使われている。

胆汁にかかわる用語 Biliary Terminology 🔊

bile	bile	胆汁
	bile acid	胆汁酸
	bile duct	胆管
	common bile duct	総胆管
bili-	**bili**ary	胆汁の
	bilirubin	ビリルビン
chole-	**chole**stasia, **chole**stasis	胆汁鬱滞
	cholecystectomy	胆囊切除（摘出）術
	cholecystokinin（CCK）	コレシストキニン*

＊　cholecyst は胆囊、kinin は動かすもの、より「胆囊を収縮させるもの」という意味で命名された。pancreozymin（パンクレオザイミン）ともいう。消化管ホルモンのひとつである。

Column

ヘパリン
　抗凝血成分のヘパリン（heparin）は、発見当時、肝臓（hepar）に多量にあると考えられたため、この名が付けられた。肝臓以外にも、小腸、筋肉、肺、脾や肥満細胞など体内に幅広く存在することが分かった今も、この名が使われている。

コレラ
　cholera（コレラ）は古くから知られていた伝染病である。スペルもまったく同じ語 "cholera" がギリシャ時代から使われていた。当時、chole（胆汁）などの体液が体外へ出てしまう病気と考えられていたことにより、この名が付いた。

b) 胆嚢と胆管 Gall Bladder and Bile Duct

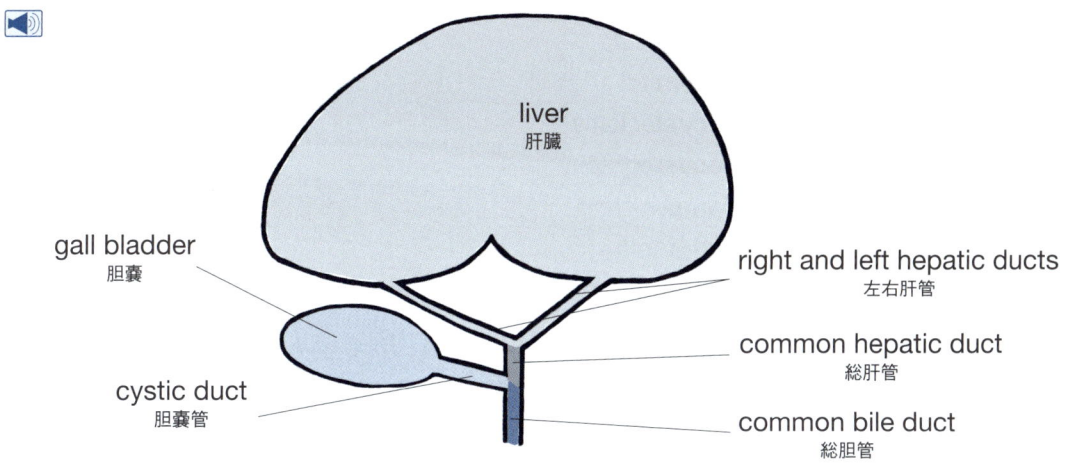

🔊

figure: liver 肝臓, gall bladder 胆嚢, cystic duct 胆嚢管, right and left hepatic ducts 左右肝管, common hepatic duct 総肝管, common bile duct 総胆管

胆嚢は二語に分けて gall bladder、一語で gallbladder とする場合があり、両方使われている。
gall（胆汁）を容れる bladder（袋）という意味である。

胆嚢 Gall Bladder

語幹	英語名詞	英語形容詞	ラテン語名詞	ギリシャ語名詞
cholecyst(o)-*	gall bladder, gallbladder, cholecystis, vesica biliaris	-	vesica fellea	-

* cholecyst(o)- =chole-（胆汁）+ cyst（袋）。

胆嚢管 Cystic Duct, Cystic Gall Duct

語幹	英語名詞	英語形容詞	ラテン語名詞	ギリシャ語名詞
-	cystic duct, cystic gall duct	-	ductus cysticus	-

胆管 Bile Duct, Bilary Duct

語幹	英語名詞	英語形容詞	ラテン語名詞	ギリシャ語名詞
cholangi(o)-[1]	bile duct, biliary duct	biliary[2]	-	-

[1] cholangi(o)- =chole-（胆汁）+ angio（管）。
[2] biliary には胆管という意味と、胆汁という意味がある。

総胆管 Common Bile Duct, Choledoch Duct

語幹	英語名詞	英語形容詞	ラテン語名詞	ギリシャ語名詞
choledoch(o)-*	common bile duct, choledoch duct	-	ductus choledochus	-

* choledoch(o)- = chole-（胆汁）+ docho（受け取る）。

胆嚢および胆管にかかわる用語 Terminology of the Gall Bladder and Bile Duct

gall bladder	gallbladder carcinoma, gall bladder cancer	胆嚢癌
cholecyst(o)-	**cholecyst**algia	胆嚢痛
	cholecystitis	胆嚢炎
	cholecystectomy	胆嚢摘出術
	cholecystography	胆嚢造影法
cholangi(o)-	**cholangi**tis	胆管炎
	cholangioma	胆管腫瘍
	cholangiocarcinoma	胆管癌
	cholangiocellular carcinoma	胆管細胞癌
choledoch(o)-	**choledoch**itis	総胆管炎

胆石 Gallstone, Biliary Calculus

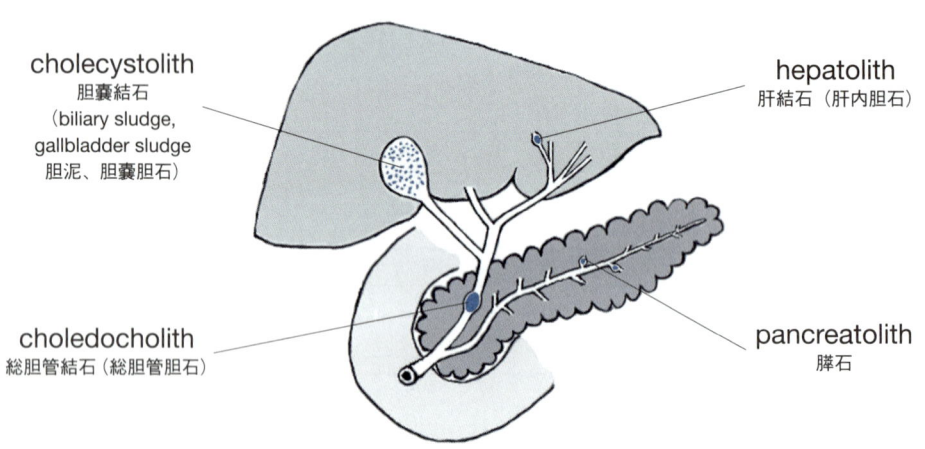

cholecystolith
胆嚢結石
（biliary sludge, gallbladder sludge
胆泥、胆嚢胆石）

hepatolith
肝結石（肝内胆石）

choledocholith
総胆管結石（総胆管胆石）

pancreatolith
膵石

胆石症 Biliary Calculosis

胆石症には、stone（石）、-lith（石）、-lithiasis（結石症）を用いる。

stone	gallstone	胆石
-lith	hepato**lith**	肝結石
	chole**lith**	胆石
	choledocho**lith**	総胆管結石
-lithiasis	hepato**lithiasis**	肝結石症
	chole**lithiasis**	胆石症
	cholecysto**lithiasis**	胆嚢に石ができる胆石症
	choledocho**lithiasis**	総胆管結石症

3 膵臓 The Pancreas

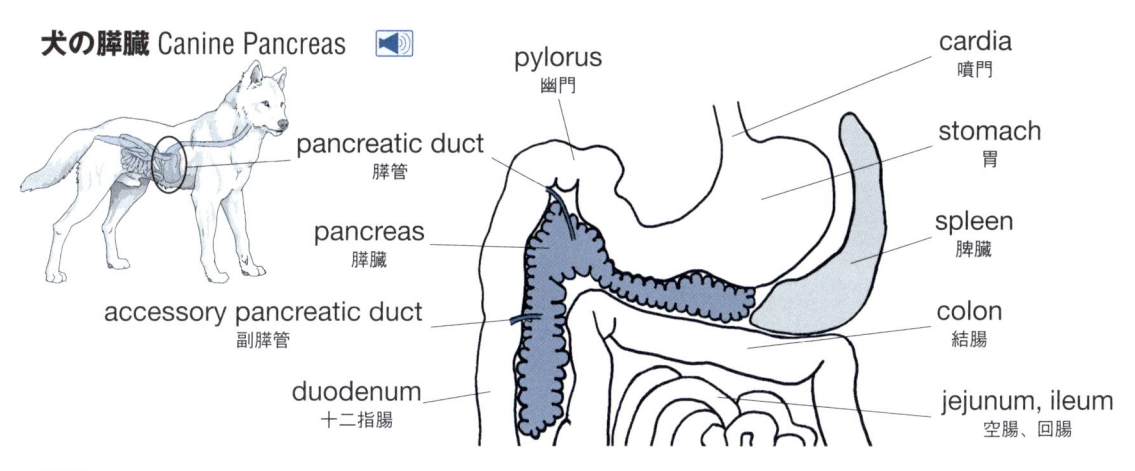

犬の膵臓 Canine Pancreas

- pylorus 幽門
- cardia 噴門
- pancreatic duct 膵管
- stomach 胃
- pancreas 膵臓
- spleen 脾臓
- accessory pancreatic duct 副膵管
- colon 結腸
- duodenum 十二指腸
- jejunum, ileum 空腸、回腸

膵臓 Pancreas

語幹	英語名詞	英語形容詞	ラテン語名詞	ギリシャ語名詞
pancreat(o)-, pancreatico-, pancreo-	pancreas	pancreatic	pancreas	pankreas

pancreas はギリシャ語の pan-（すべて）+ kreas（肉様物質）に由来する。確かに見た目、やや赤みがかって肉様と言えなくもない。

膵臓にかかわる用語 Pancreatic Terminology

pancreatic	pancreatic vein		膵静脈
	pancreatic cancer		膵臓癌
pancreat(o)-, pancreo-	**pancreat**algia		膵（臓）痛
	pancreatitis 膵炎	acute **pancreat**itis	急性膵炎
		chronic **pancreat**itis	慢性膵炎
	pancreatolith, **pancreo**lith (=pancreatic calculus)		膵石
	pancreatolithiasis		膵石症
	pancreatectomy		膵切開術
	pancreozymin		パンクレオザイミン*
pancreatico-	**pancreatico**splenic lymph nodes		膵脾リンパ節
	pancreaticoduodenal		膵十二指腸の
その他	diabetes mellitus (DM)		糖尿病

＊ 消化管ホルモンのひとつ。コレシストキニンに同じ。

膵臓外分泌部、内分泌部 Exocrine and Endocrine Pancreas

exocrine part of pancreas		膵外分泌部
endocrine part of pancreas 膵内分泌部	pancreatic islets (=islets of Langerhans)	膵島（ランゲルハンス島）
	alpha cell (glucagon)	アルファ細胞（グルカゴン）
	beta cell (insulin)	ベータ細胞（インスリン）
	delta cell (somatostatin)	デルタ細胞（ソマトスタチン）
	PP cell (pancreatic polypeptide)	PP細胞（膵ポリペプチド）

Exercises 7

問1．下の語群から（　）に適当なものを選びなさい。

1. 牛の第一胃は（①　　　）、第二胃は（②　　　）という。reticulum という単語は、細胞小器官である小胞体（③　　　）を表し、ER と略される。粗面小胞体は（④　　　）、滑面小胞体は（⑤　　　）である。

    a) rough ER/rER　b) smooth ER/sER　c) rumen　d) reticulum
    e) endoplasmic reticulum

2. 牛の心臓は、横隔膜（①　　　）を隔てて、第二胃（②　　　）と接している。牛が誤って、釘のようなとがった異物を飲み込んだ場合、その釘が胃壁、さらに横隔膜を貫通することがある。甚だしい場合は心膜（③　　　）を傷付け、炎症（④　　　）をおこすこともある。
    第二胃の炎症は（⑤　　　）、横隔膜の炎症は inflammation of diaphragm（diaphragmitis）、心膜の炎症は（⑥　　　）である。釘などの異物による心膜炎を「牛の創傷性心膜炎」（⑦　　　）と言う。

    a) reticulum　b) reticulitis　c) diaphragm　d) inflammation
    e) bovine traumatic pericarditis　f) pericardium　g) pericarditis

問2．次の用語の左右を正しく結びなさい。

1. ①肝腫大　　　　　　　・　　　　　　・a) fatty liver, hepatic steatosis
   ②黄疸　　　　　　　　・　　　　　　・b) hepatomegaly
   ③肝硬変　　　　　　　・　　　　　　・c) liver cirrhosis
   ④脂肪肝　　　　　　　・　　　　　　・d) jaundice, icterus
   ⑤肝性昏睡　　　　　　・　　　　　　・e) hepatic coma

2. ①門脈　　　　　　　　・　　　　　　・a) central vein
   ②後大静脈　　　　　　・　　　　　　・b) interlobular connective tissue, fibrous capsule of liver
   ③肝小葉　　　　　　　・　　　　　　・c) hepatic lobule
   ④中心静脈　　　　　　・　　　　　　・d) posterior vena cava
   ⑤小葉間結合組織　　　・　　　　　　・e) portal vein

3. ①膵島　　　　　　　　・　　　　　　・a) endocrine
   ②腺房　　　　　　　　・　　　　　　・b) islets
   ③外分泌　　　　　　　・　　　　　　・c) acinus
   ④内分泌　　　　　　　・　　　　　　・d) pancreatic duct
   ⑤膵管　　　　　　　　・　　　　　　・e) exocrine

Exercises7の答え

問1．1. ① c)　② d)　③ e)　④ a)　⑤ b)　　2. ① c)　② a)　③ f)　④ d)　⑤ b)　⑥ g)　⑦ e)
問2．1. ① b)　② d)　③ c)　④ a)　⑤ e)　　2. ① e)　② d)　③ c)　④ a)　⑤ b)
　　　3. ① b)　② c)　③ e)　④ a)　⑤ d)

Chapter 9

尿生殖器系とその疾患
The Urogenital System and its Disorders

1 泌尿器 The Urinary Organs

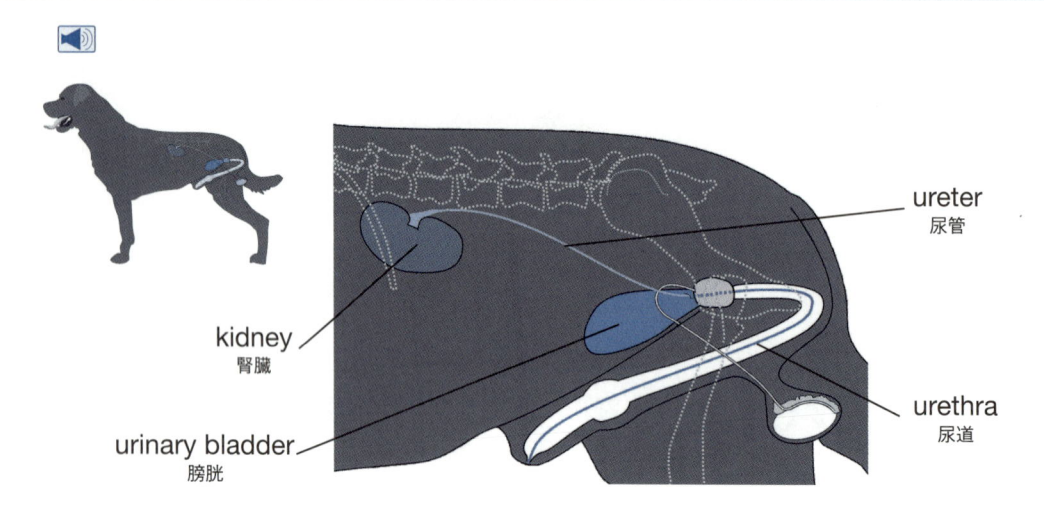

- ureter 尿管
- kidney 腎臓
- urinary bladder 膀胱
- urethra 尿道

a) 腎臓 Kidney

語幹	英語名詞	英語形容詞	ラテン語名詞	ギリシャ語名詞
nephr(o)-, ren(o)-	kidney	renal	ren	nephros

ギリシャ語の"Nephros"から、ネフロン（腎単位）やネフローゼが派生した。

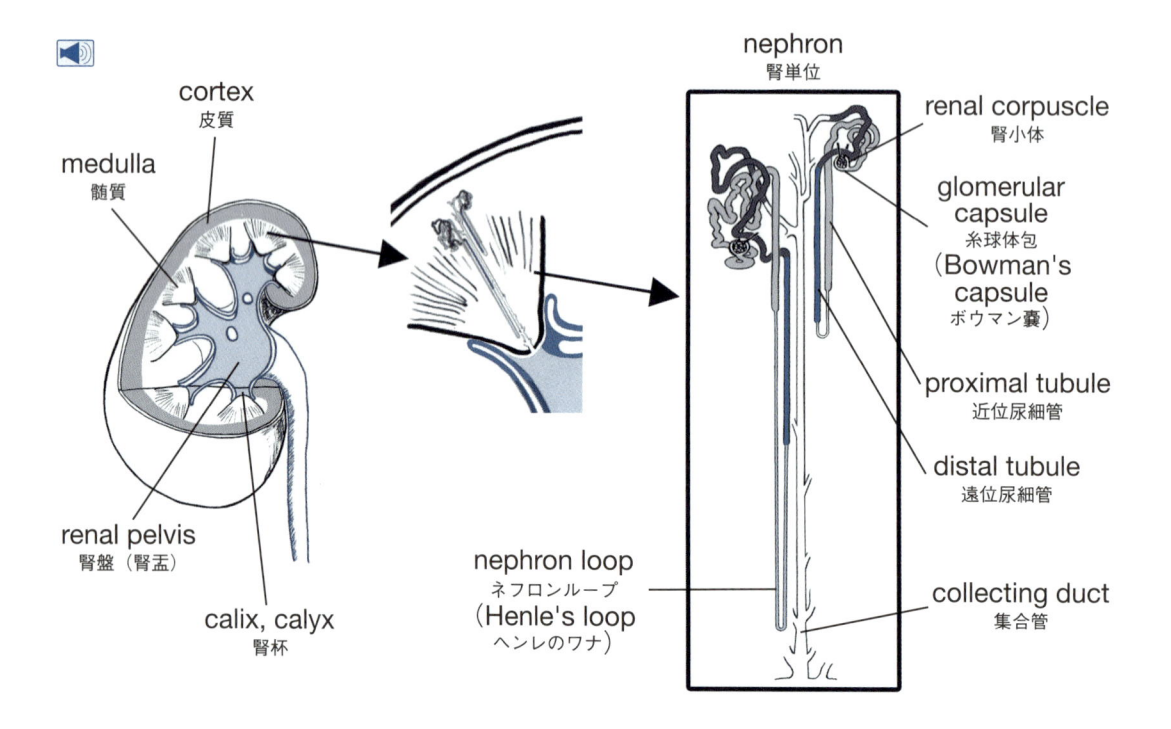

- cortex 皮質
- medulla 髄質
- renal pelvis 腎盤（腎盂）
- calix, calyx 腎杯
- nephron 腎単位
- renal corpuscle 腎小体
- glomerular capsule 糸球体包 （Bowman's capsule ボウマン嚢）
- proximal tubule 近位尿細管
- distal tubule 遠位尿細管
- nephron loop ネフロンループ （Henle's loop ヘンレのワナ）
- collecting duct 集合管

122

腎臓にかかわる用語 Renal Terminology 🔊

kidney	kidney cancer		腎臓癌
	kidney transplant		腎臓移植
	cystic kidney		嚢胞腎
renal	renal failure 腎不全	acute renal failure（ARF）	急性腎不全
		chronic renal failure（CRF）	慢性腎不全
	renal calculus（=**reno**lith）		腎臓結石、腎石症
ren(o)-	**reno**megaly		腎肥大［症］
	renin		レニン（レニン・アンギオテンシン系）
	ad**ren**al gland		副腎。adは近傍、の意。suprarenal glandともいう。adrenalineアドレナリンの名もこれに由来。
nephr(o)-	**nephr**algia		腎臓痛
	nephrectomy		腎臓摘出
	nephritis 腎炎	glomerulo-**nephr**itis	糸球体腎炎
		tubulo- interstitial **nephr**itis	尿細管間質性腎炎
		embolic **nephr**itis	塞栓性腎炎（化膿性腎炎または腎膿瘍）
	nephrosis		ネフローゼ
	nephrosclerosis		腎硬化症
	hydro**nephro**sis		水腎症
pyel(o)-* 腎盂	**pyel**itis		腎盂炎
	pyelonephritis		腎盂腎炎
	retrograde **pyelo**graphy		逆行性腎盂造影

＊ ギリシャ語のpyelos（水を入れる浅くて丸い容器）より。

b) 尿管 Ureter

語幹	英語名詞	英語形容詞	ラテン語名詞	ギリシャ語名詞
ureter(o)-	ureter	ureteral	ureter	ourētēr

尿管にかかわる用語 Ureteral Terminology 🔊

ureter(o)-	**ureter**algia	尿管痛
	ureteritis	尿管炎
	ureterocystostomy	尿管膀胱吻合［術］

c) 膀胱 Urinary Bladder

語幹	英語名詞	英語形容詞	ラテン語名詞	ギリシャ語名詞
cyst(o)-, cysti-	urinary bladder	cystic, vesical	vesica urinaria	kystis

　　ギリシャ語の kystis（袋）より、「尿を蓄える袋」という意味。ただし、ギリシャ語の kystis には、袋、嚢、という意味しかないので、他の袋状の組織、例えば胆嚢管、嚢胞も同じ語幹を持つ。

膀胱にかかわる用語 Urinary Bladder Terminology 🔊

bladder	paralysis of urinary bladder	膀胱麻痺
	rupture of bladder	膀胱破裂
vesical	vesical calculus（=bladder stones, cystolith）	膀胱結石症
cyst(o)-	**cyst**algia	膀胱痛
	cystitis	膀胱炎
	cystogram	膀胱造影像
	retrograde **cysto**graphy	逆行性膀胱造影

d) 尿道 Urethra

語幹	英語名詞	英語形容詞	ラテン語名詞	ギリシャ語名詞
urethr(o)-	urethra	urethral	urethera	-

尿道にかかわる用語 Urethral Terminology 🔊

urethr(o)-	**urethr**algia, **urethro**dynia	尿道痛
	urethritis	尿道炎

> **Column 遊走する臓器たち**
>
> 　　牛などの反すう動物では、左の腎臓の位置は可動性に富む（flexible）。左腹腔の大部分を占める第一胃（rumen）が草でいっぱいに膨らんでいると、左腎は押されて前後あるいは右腹腔へ動く。こういった可動性に富む腎臓を遊走腎（floating kidney）と呼び、反すう動物ではこれが正常な姿である。
>
> 　　人間だと、こうはいかない。人体の医学用語で遊走腎というと、れっきとした病気である。人間は直立しているため腎臓の固定が弱まると、腎臓は重力で下へ下がるので、遊走腎のことを腎下垂症（nephroptosis）とも言う。「遊走」という言葉の付く人体医学用語は、他に遊走心（cor mobile）、遊走肝（wandering liver）、遊走脾（floating spleen）などがある。いろいろな臓器が遊走するのである！

e) 尿 Urine

語幹	英語名詞	英語形容詞	ラテン語名詞	ギリシャ語名詞
ur (o)-	urine	urinary	urina	-

尿にかかわる用語 Urine-related Terminology 🔊

ur (o)-	**ur**ine／**ur**inate	尿／排尿する
	uremia	尿毒症
	hypo**ur**esis	乏尿症
	retrograde **uro**graphy	逆行性尿路造影
	excretory **uro**graphy	排泄性尿路造影
urinary	urinary stone	尿石、尿路結石
	urinary retention	尿閉
-uria*	poly**uria**	多尿
	pollaki**uria** （=frequent urination）	頻尿
	olig**uria**	乏尿、尿量減少
	an**uria**	無尿〔症〕
	hemat**uria**	血尿〔症〕
	py**uria**	膿尿〔症〕

＊ -uria は尿の状態を表す接尾辞。

尿路の結石いろいろ Urinary Calculus

尿路には結石ができやすい。一般に結石名は、「臓器名＋stone」で表す。医学用語は calculus(sg.)、calculi(pl.)を含むか、あるいは -lith という接尾辞を持つ。尿路結石症は urolithiasis である。

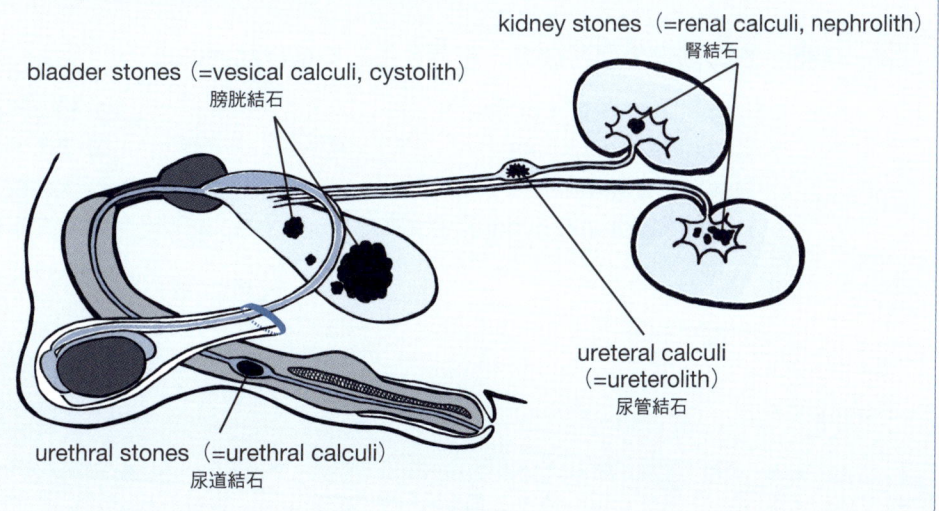

kidney stones （=renal calculi, nephrolith）
腎結石

bladder stones （=vesical calculi, cystolith）
膀胱結石

ureteral calculi
（=ureterolith）
尿管結石

urethral stones （=urethral calculi）
尿道結石

Exercises 8

問1．臓器と炎症名を正しく結びなさい。

① kidney　　　　　　・　　　・ a) cystitis
② ureter　　　　　　・　　　・ b) nephritis
③ urinary bladder　・　　　・ c) ureteritis
④ urethra　　　　　・　　　・ d) urethritis

問2．次の用語の左右を正しく結びなさい。

①腎結石　　　・　　　　・ a) ureteral stones, ureterolith, ureteral calculi
②尿管結石　　・　　　　・ b) kidney stones, nephrolith
③膀胱結石　　・　　　　・ c) bladder stones, cystolith
④尿道結石　　・　　　　・ d) urethral stones, urethral calculi

問3．下の語群から（　）に適当なものを選びなさい。

腎不全（①　　　　）が進行すると、腎機能が低下して尿（②　　　　）として体外に排泄（③　　　　）されるべき毒素が十分に排泄されなくなる。その結果、毒素（④　　　　）などが体内に蓄積されるため、全身の臓器（⑤　　　　）にさまざまな障害を与える。これを尿毒症（⑥　　　　）という。

a) renal failure　　b) uremia　　c) organs　　d) toxins　　e) urine
f) excretion

問4．次の尿のトラブルの左右を正しく結びなさい。

① polyuria　　　　　　　　　　　　　・　　・ a) 血尿
② frequent urination, pollakiuria　・　　・ b) 無尿
③ oliguria　　　　　　　　　　　　　・　　・ c) 乏尿
④ anuria　　　　　　　　　　　　　　・　　・ d) 頻尿
⑤ hematuria　　　　　　　　　　　　・　　・ e) 多尿
⑥ pyuria　　　　　　　　　　　　　　・　　・ f) 膿尿

問5．次の動物の尿石症の、左右を正しく結びなさい。

① urolithiasis of cattle　　　　　　　・　　・ a) 猫の尿石症
② canine urolithiasis　　　　　　　　・　　・ b) 猫の泌尿器症候群
③ feline urolithiasis　　　　　　　　・　　・ c) 犬の尿石症
④ feline urologic syndrome（FUS）・　　・ d) 牛の尿石症

Exercises 8 の答え

問1．① b)　② c)　③ a)　④ d)　問2．① b)　② a)　③ c)　④ d)　問3．① a)　② e)　③ f)　④ d)　⑤ c)　⑥ b)
問4．① e)　② d)　③ c)　④ b)　⑤ a)　⑥ f)　問5．① d)　② c)　③ a)　④ b)

2 雌性生殖器 The Female Genital Organs

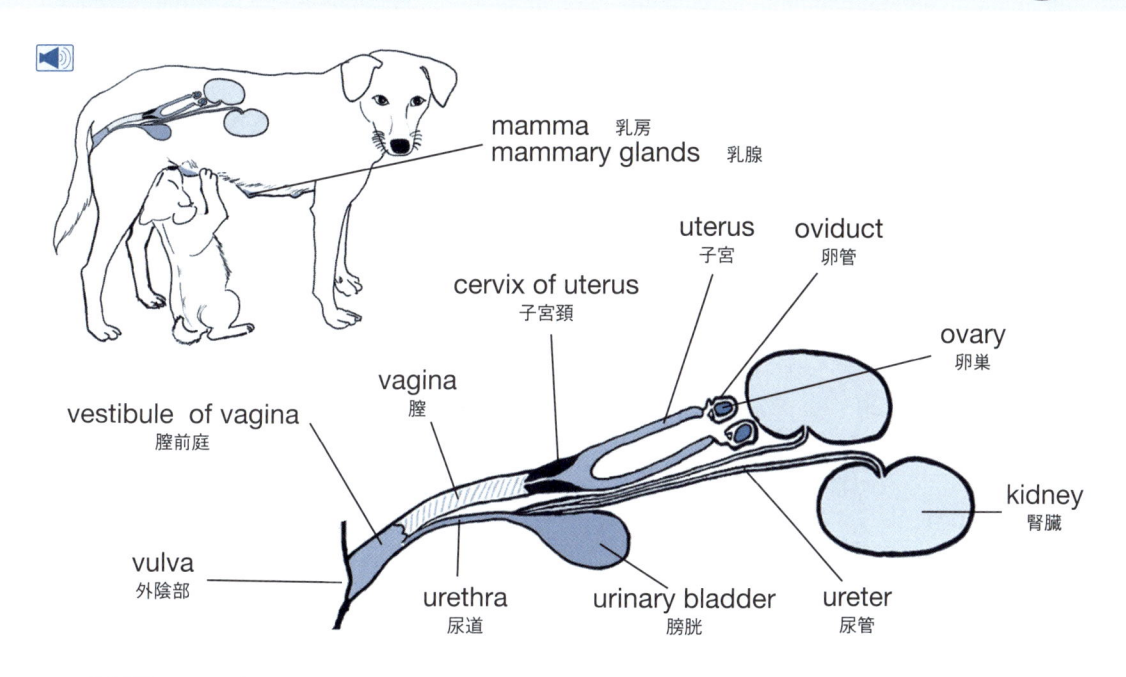

（図：犬の雌性生殖器）
- mamma 乳房
- mammary glands 乳腺
- uterus 子宮
- cervix of uterus 子宮頸
- oviduct 卵管
- ovary 卵巣
- vagina 膣
- vestibule of vagina 膣前庭
- kidney 腎臓
- vulva 外陰部
- urethra 尿道
- urinary bladder 膀胱
- ureter 尿管

a) 卵巣 Ovary

語幹	英語名詞	英語形容詞	ラテン語名詞	ギリシャ語名詞
ovari(o)-	ovary	ovarian	ovarium	-
oo-, oophor(o)-	-	-	oöphoron	öion

　卵巣に関する接頭辞には、ovari(o)- と oophor(o)- がある。前者はラテン語で卵を表す ovum(sg.)、ova(pl.) に由来する。この ovum という単語は「卵子」として、現在医学用語として使われている。後者の oophor(o)- はギリシャ語の öio（卵）に由来し、これからラテン語の oöphoron（卵巣）となった。

卵巣にかかわる用語 Ovarian Terminology

語幹	用語	和訳
ovari(o)-, oophor(o)-	**ovari**algia, **oophor**algia	卵巣痛
	ovaritis, **oophor**itis	卵巣炎
	ovarioncus, **oophor**oma	卵巣腫
	ovariohysterectomy（OHE）	卵巣子宮切除術*
	ovariectomy, **oophor**ectomy	卵巣摘出〔術〕
ovarian	ovarian cyst	卵巣嚢腫
	ovarian adhesion	卵巣癒着
その他	ovulation failure	排卵障害
	luteal hypoplasia	黄体形成不全
	retained corpus luteum, retention of corpus luteum	黄体遺残

＊ 避妊法に用いられる手術法。

[図: 卵巣と子宮における卵子の動き]
- implantation 着床
- placenta 胎盤
- endometrium 子宮内膜
- corpus luteum 黄体
- ovulation 排卵
- fertilization 受精

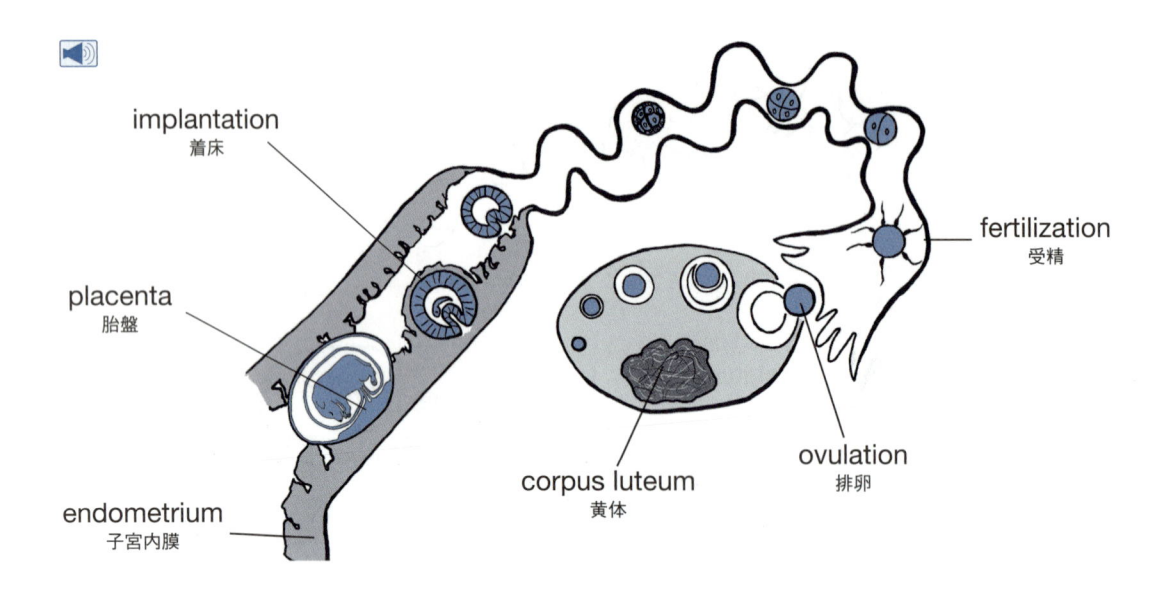

卵子（Ovum）のドラマ

すなわち primordial ovarian follicle（原始卵胞）→ primary ovarian follicle（一次卵胞）→ secondary ovarian follicle（二次卵胞）→ mature ovarian follicle, Graafian follicle（成熟卵胞）→ ovulation（排卵）[卵巣において] → fertilization（受精）[卵管膨大部において] → cleavage（卵割）→ implantation（着床）[子宮角の子宮内膜内] となる。

b) 卵管 Oviduct

語幹	英語名詞	英語形容詞	ラテン語名詞	ギリシャ語名詞
salping(o)-, tub(o)-	oviduct	salpingian	tuba uterina	salpinx*

卵管は日本語も英語も「卵巣（卵）」に由来する言葉だが、ラテン語は"tuba uterina"と、子宮に由来しているところが興味深い。英語形容詞の salpingian は、もともと「ラッパの」という意味で、「耳管の」も同じく salpingian を用いる。卵管も耳管もトランペットのように一端が広がっているからだろう。oviductの他に、fallopian tube という少し古い表現もある。

＊ ラッパという意味。

卵管にかかわる用語 Oviduct Terminology

salping(o)-	**salping**itis	卵管炎
	salpingoma	卵管腫瘍
	salpingectomy（=**tub**ectomy）	卵管摘出〔術〕
salpinx	hydrosalpinx	卵管水腫
	pyosalpinx	卵管蓄膿症
tub(o)-	**tub**o-ovaritis（=**salpingo**-oophoritis）	卵管卵巣炎

c) 子宮 Uterus

語幹	英語名詞	英語形容詞	ラテン語名詞	ギリシャ語名詞
uter(o)-, uteri-, hyster(o)-, metr(o)-, -metra	uterus(sg.), uteri(pl.), womb	uterine	uterus	hysteria, mētra

hysteria からはヒステリーという語も派生した。昔はヒステリーは女性特有とみなされていたためである。また、womb は主に、妊娠子宮を意味する。

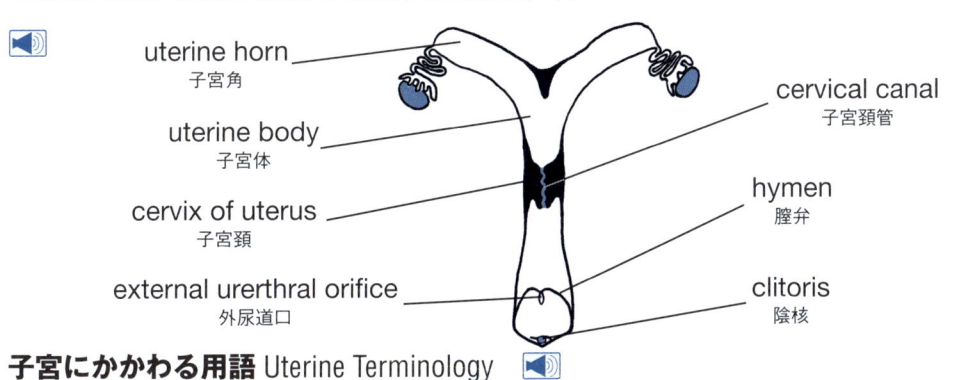

- uterine horn 子宮角
- uterine body 子宮体
- cervix of uterus 子宮頚
- external urerthral orifice 外尿道口
- cervical canal 子宮頚管
- hymen 膣弁
- clitoris 陰核

子宮にかかわる用語 Uterine Terminology

語幹	用語	和訳
uter(o)-, uteri-	**uter**itis (=**metr**itis)	子宮炎
	extra**uter**ine pregnancy (=ectopic pregnancy)	子宮外妊娠
uterine	uterine torsion	子宮捻転
	uterine hernia	子宮ヘルニア
	uterine prolapse	子宮脱
hyster(o)-	**hyster**algia, **hyster**odynia	子宮痛
	hysterogram	子宮造影図
	hysterography	子宮造影法
	hysteroscope	子宮鏡（子宮用内視鏡）
	hysterectomy	子宮摘出〔術〕
metr(o)-	endo**metr**itis	子宮内膜炎
	peri**metr**itis	子宮外膜炎
	myo**metr**itis	子宮筋炎
-metra	pyo**metra**	子宮蓄膿症
	muco**metra**	子宮粘液症
	hydro**metra**	子宮水腫
cervical canal 子宮頚管	cervical stenosis	頚管狭窄
	cervicitis	頚管炎

Column 結婚の神様

Hymen は人間では処女膜、動物では膣弁と訳される。偶然だが、古代ギリシャの結婚の神様は Hymen という。とりわけ花婿の家に向かう花嫁の行列によって歌われた賛歌の神様である。

d) 腟 Vagina

語幹	英語名詞	英語形容詞	ラテン語名詞	ギリシャ語名詞
vagin(o)-, colp(o)-	vagina	vaginal	vagina	kolpos

vagina は ラテン語で「刀剣の鞘」、vaginate は「鞘のある」という意味。語幹 colpo- はギリシャ語の kolpos「空所の、中空の」に由来する。

腟にかかわる用語 Vaginal Terminology 🔊

語幹	用語	和訳
vagin(o)-	**vagino**dynia（ =**colp**algia, **colpo**dynia）	腟痛
	vaginitis（ =**colp**itis）	腟炎
	bacterial **vagino**sis（BV）	細菌性腟症
	vaginohysterectomy （=vaginal hysterectomy）	腟子宮摘出［術］
vaginal	vaginal plug	腟栓
	vaginal stenosis	腟狭窄
	vaginal cyst	腟嚢胞
	vaginal prolapse	腟脱
vagina	urovagina	尿腟
	pneumovagina	気腟
colp(o)-	**colp**ectomy	腟切除術
	colposcope	腟鏡
	colpatresia	腟閉鎖［病］

e) 乳房 Mamma, Udder

語幹	英語名詞	英語形容詞	ラテン語名詞	ギリシャ語名詞
mamm(o)-, mast(o)-	mamma, udder	mammal	mamma*	mastos

mamma は乳房。udder は牛や山羊など有蹄類の袋状に垂れ下がった乳房を意味する。乳房炎は mammaritis ではなく、mastitis なので注意しよう。udder の語源は「サンスクリット語の ūdhar にさかのぼる」という説と、「西ドイツ起源で古英語の ūder に由来する」という説がある。

* ラテン語の mamma には「乳房」と「お母さん」というふたつの意味がある。この mamma から mammal（哺乳類）が派生した。

乳房にかかわる用語 Mammary Terminology 🔊

語幹	用語	和訳
mamm(o)-	**mamm**algia	乳房痛
	mammography	乳房撮影［法］*
	mammectomy （=**mast**ectomy）	乳房切除［術］
mammary	mammary tumor	乳腺腫瘍
	mammary gland	乳腺
	mammary tumor, mammary adenocarcinoma （=breast cancer）	乳癌
mast(o)-	**mast**algia, **masto**dynia	乳房痛
	mastitis	乳房炎
	mastadenoma	乳房腺腫

* 乳房のX線、超音波、MRI などによる画像検査。

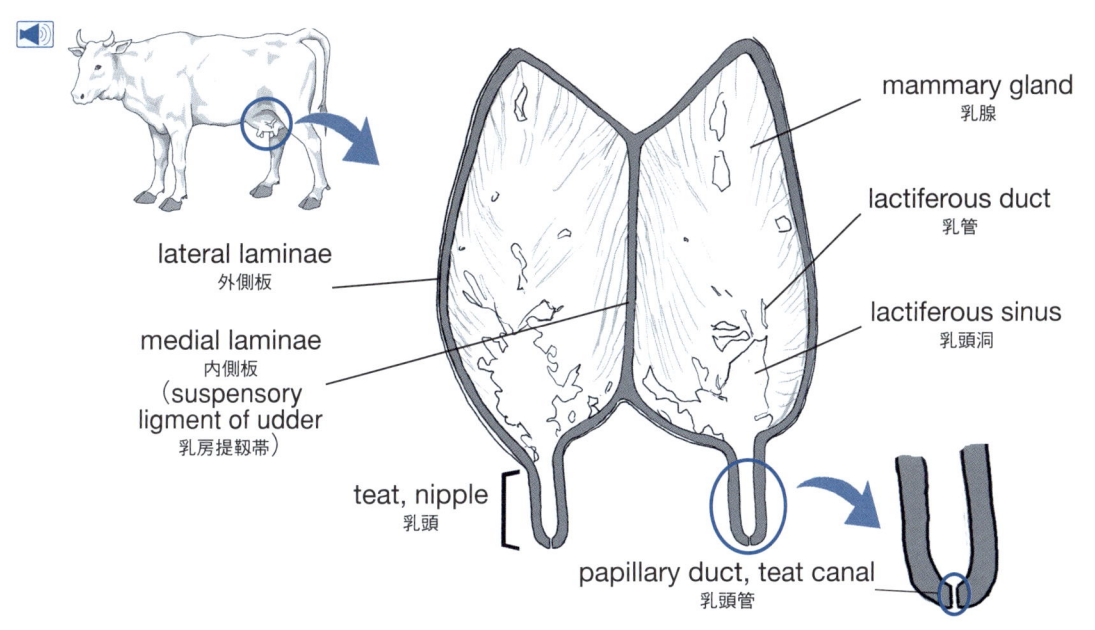

f) 乳汁 Milk

語幹	英語名詞	英語形容詞	ラテン語名詞	ギリシャ語名詞
lact(o)-, lacti-, galact(o)-	milk	lactic	lac, lactis*	gala, galaktos*

　Milk はゲルマン語に由来。ドイツ語は Milch 。

* 属格（＝英語でいう所有格）。

乳汁にかかわる用語 Lactation Terminology

milk	milky	乳白色の、ミルクのような
	milking	乳搾り
	colostrum*, foremilk, neogala	初乳
	milk line	乳腺堤（乳房の列）
	milk fever	乳熱
lact(o)-, lacti-	**lact**ation	乳汁分泌
	lactiferous	乳汁分泌性の
	lactose	乳糖
lactic	lactic acid	乳酸
galact(o)-	**galacto**poietic	乳汁産生の
	galactose	ガラクトース
	galactosamine	ガラクトサミン
	hyper**galact**ia	乳汁分泌過多症

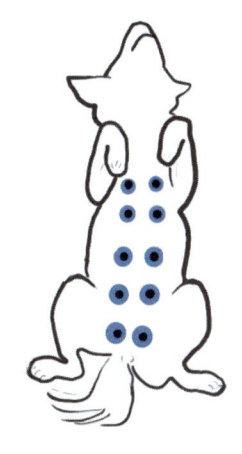

犬のMilk line
4～6対の乳房が腋窩から鼡径部にかけて並ぶ。

* ラテン語の colostra（初乳）に由来し、英語に取り込まれたのは16世紀のことである。それまで使われていた beestings や、green milk という言葉は colostra に取って代わられ、もはや使われなくなった。19世紀に語尾が変化し、現在の形になった。

3 雄性生殖器 The Male Genital Organs

a) 精巣 Testis 🔊

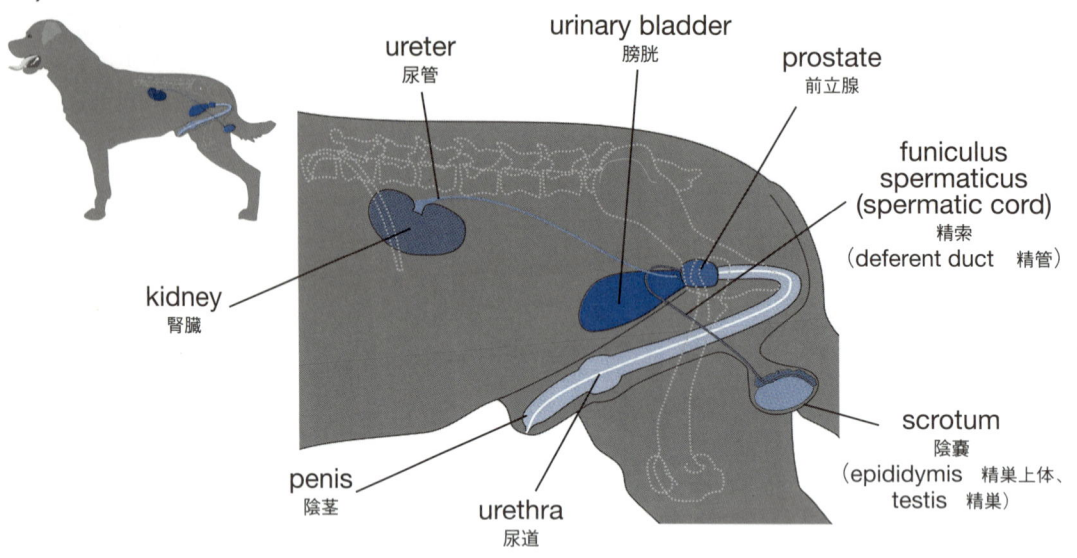

- ureter 尿管
- urinary bladder 膀胱
- prostate 前立腺
- funiculus spermaticus (spermatic cord) 精索 (deferent duct 精管)
- kidney 腎臓
- scrotum 陰嚢 (epididymis 精巣上体、testis 精巣)
- penis 陰茎
- urethra 尿道

語幹	英語名詞	英語形容詞	ラテン語名詞	ギリシャ語名詞
test-, orchi-	testis(sg.), testes(pl.),orchis	testicular	testis, orchis	orkhis

　一般用語では、睾丸は testicle である。-cle は、小さなものに付ける接尾辞。他に didymus という言い方もある。ギリシャ語の didymos（双子）に由来する。これに上を表す epi- が付いて、epididymus（精巣上体）となった。

精巣にかかわる用語 Testicular Terminology 🔊

test-	**test**algia （=**orchi**algia, **orchio**dynia）	精巣痛、睾丸痛
	testitis （=didymitis, **orchi**tis）	精巣炎、睾丸炎
testicular	testicular tumor （=**orchi**oncus）	精巣（睾丸）腫瘍
	testicular hypoplasia	精巣発育不全
orchi-	crypt**orchi**sm （=undescended testis, retained testis）	停滞睾丸[*1]
	orchiopexy	精巣（睾丸）固定術[*2]
その他	seminoma	セミノーマ（精上皮腫）
	interstitial cell tumor	間質細胞腫
	Leydig cell tumor	ライディッヒ細胞腫
	Sertoli cell tumor	セルトリ細胞腫
	cremaster muscle[*3]	精巣挙筋

[*1] cryptは「隠れた」を意味し、精巣下降（descensus testis）がうまくいかなかった状態をいう。
[*2] 下降していない精巣を降ろし、陰嚢にはめ込む手術のこと。
[*3] cremaster muscle（精巣挙筋）は、ギリシャ語でサスペンダーを意味する kremaster に由来する。つまり精巣をサスペンダーのように吊り上げている筋肉、という意味である。

b) 雄性副生殖腺 Male Accessory Genital Glands 🔊

馬の雄性生殖器
Equine Male Genital Organs

精嚢腺（seminal vesicle）の semi- は精液（semen）の語幹である。
実は分泌腺であるが、作られた精液を貯蔵するところと誤解されていたため、この名が付いた。犬や猫にはない。馬やウサギは長い袋状で腺、という言葉を取って、「精嚢」と呼ぶ。

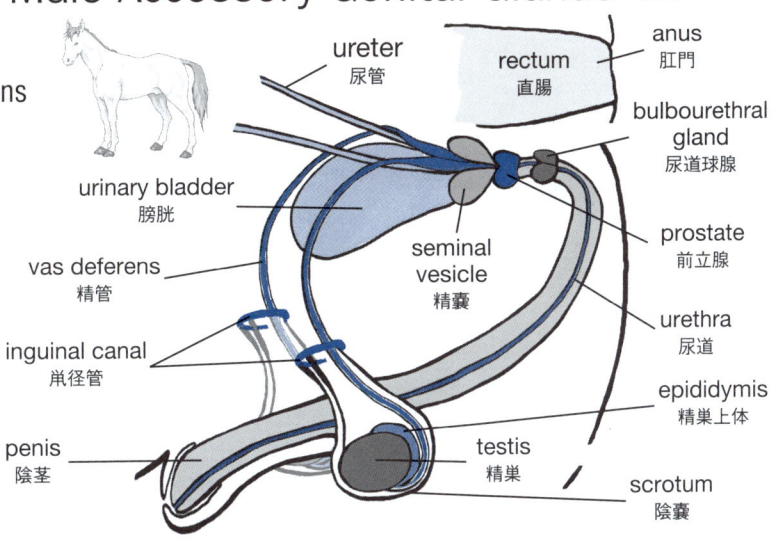

図：ureter 尿管／rectum 直腸／anus 肛門／bulbourethral gland 尿道球腺／prostate 前立腺／urethra 尿道／epididymis 精巣上体／scrotum 陰嚢／testis 精巣／penis 陰茎／inguinal canal 鼡径管／vas deferens 精管／urinary bladder 膀胱／seminal vesicle 精嚢

精巣上体 Epididymis

語幹	英語名詞	英語形容詞	ラテン語名詞	ギリシャ語名詞
epididym(o)- orchid-	epididymis (sg.), epididymides (pl.)	epididymal	epididymitis	epididymitis*

* epididymis = epi（上）＋ didymis（精巣）。

精管 Deferent Duct

語幹	英語名詞	英語形容詞	ラテン語名詞	ギリシャ語名詞
defero-, deferent-	deferent duct	-*¹ (deferential)	ductus deferens*²	-

*1 英語名詞が2語（以上）の場合は形容詞の代わりに of でつなぐ、すなわち「精管の」は - of seminal vesicle である。
*2 defero- は「取り去る、運び出す」の意味。

精嚢腺 Seminal Vesicle

語幹	英語名詞	英語形容詞	ラテン語名詞	ギリシャ語名詞
semi-*	seminal vesicle	-	glandulae vesicularis	-

* 精管を表す単数形は、他に ductus deferens, seminal duct, spermatic duct, vas deferens、複数形は deferent ducts, seminal ducts, spermatic ducts, vasa deferentia などがある。

前立腺 Prostate Gland

語幹	英語名詞	英語形容詞	ラテン語名詞	ギリシャ語名詞
prostat-	prostate gland, prostate, prostata	prostatic	prostata	prostatēs

尿道球腺 Bulbourethral Gland

語幹	英語名詞	英語形容詞	ラテン語名詞	ギリシャ語名詞
-	bulbourethral gland	-	glandula bulbourethralis	-

雄性副生殖腺にかかわる用語 Terminology of Male Accessory Genital Glands 🔊

epididym(o)-	**epididym**itis	精巣上体炎
deferent-	**deferent**itis（=vasitis）	精管炎
	vasectomy*	精管切除
semi-	**semi**no-vesiculitis（=seminal vesiculitis）	精囊腺炎
prostat-	**prostat**algia, **prostat**odynia	前立腺痛
	prostatitis	前立腺炎
	prostatomegaly, prostatic hypertrophy	前立腺肥大
bulbourethral gland	bulbitis	尿道臭腺炎

* vasectomy = vas deferens + ectomy。

上記以外の雄性器官 Other Male Genital Organs

語幹	英語名詞	英語形容詞	ラテン語名詞	ギリシャ語名詞	和訳
defero-, deferent-	funiculus spermaticus*1, spermatic cord	-	funiculus	-	精索
scrot(o)-*2, osche(o)-	scrotum(sg), scrota(pl.)	scrotal, oscheal	scrotum	osche	陰嚢
phall(o)-	penis*3(sg.), penes(pl.)	penile	penis	phallos	陰茎

*1 精索は精管や血管を含む複合構造。精索炎は funiculitis。
*2 陰囊炎は scrotitis。
*3 penis の英語の発音はピーナスで、スペル通りではない。この語はもともとはラテン語の「尾」という意味。

> **Column** 目撃者と蘭
>
> 精巣の英語にはラテン語由来の testis と、ギリシャ語由来の orchis のふたつがある。ラテン語の testis には「目撃者」という意味がある。
> ギリシャ語の orkhis はもともと「睾丸」という意味であるが、この言葉は、蘭（orchis, orchid）の語源ともなった。野性の蘭の根は、ふたつの塊状に肥大するものがある。その姿が精巣に似ていることから、蘭は「睾丸」という名前をもらったのである。しかも紀元1世紀のプリニウスの博物誌には、「蘭の根を水にいれて飲むと性欲が亢進される」と、書かれているそうだ。ただし、この博物誌に書かれている多くの事柄の真偽のほどは疑問があるので性欲亢進の方もはなはだ怪しいが。

c) 精子と精液 Sperm and Semen

語幹	英語名詞	英語形容詞	ラテン語名詞	ギリシャ語名詞
sperm-, spermat(o)-	sperm*	spermatic	spermium, spermatozoon	sperma
semi-	semen	seminal	semen	-

　精子のことを昔は「精虫」とも言った。ギリシャ語の zōion は動物という意味で、ここから zoo（動物園）や zoology（動物学）が派生した。この zōion がラテン語に取り入れられ、ラテン語の精子は spermatozoon となった。昔、zoon を日本語にするとき、「虫」と訳したわけである。

　精液の semen はもともとはラテン語で「種」という意味で、serere（種を撒く）に由来する。ここから派生して英語の inseminate という動詞には「種をまく、植えつける、繁殖させる」、名詞の semination は「播種」、seminality は「萌芽、芽生え」という意味がある。

　精子は動物に、精液は種（植物）に由来する、本来は別々の言葉であったが、日本語に訳すとき、両方「精」の字を付けて、ひとくくりにしたわけである。

＊　古い言い方に spermatozoon がある。

精子および精液にかかわる用語 Terminology of the Sperm and Semen 🔊

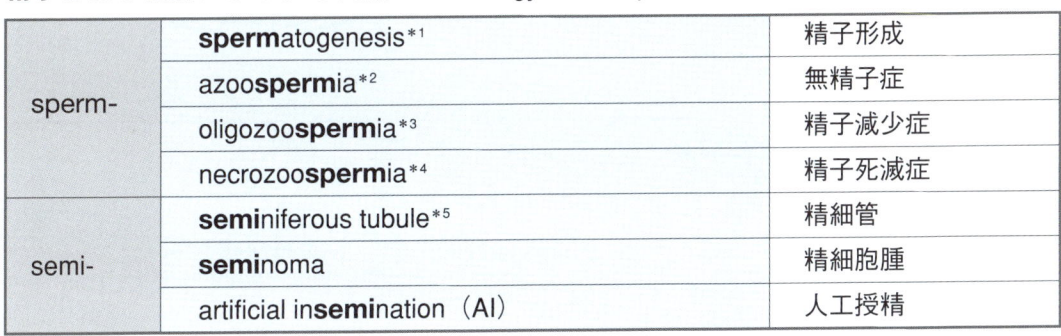

sperm-	**sperm**atogenesis*¹	精子形成
	azoo**sperm**ia*²	無精子症
	oligozoo**sperm**ia*³	精子減少症
	necrozoo**sperm**ia*⁴	精子死滅症
semi-	**semi**niferous tubule*⁵	精細管
	seminoma	精細胞腫
	artificial in**semi**nation（AI）	人工授精

＊1　spermatogenesis = sperm ＋ genesis。
＊2　azoospermia = a-（否定）＋ zoo ＋ sperm ＋ -ia（病気）。azoospermia（無精子症）と紛らわしいものに aspermia（無精液症）がある。別々の症状なので注意。
＊3　oligozoospermia = oligo（small number）＋ zoo ＋ sperm ＋ -ia。
＊4　necrozoospermia = necro（death）＋ zoo ＋ sperm ＋ -ia。
＊5　seminiferousは「種子を生じる、精子を生じる」の意味。

4 臨床繁殖学
Clinical Reproduction / Theriogenology

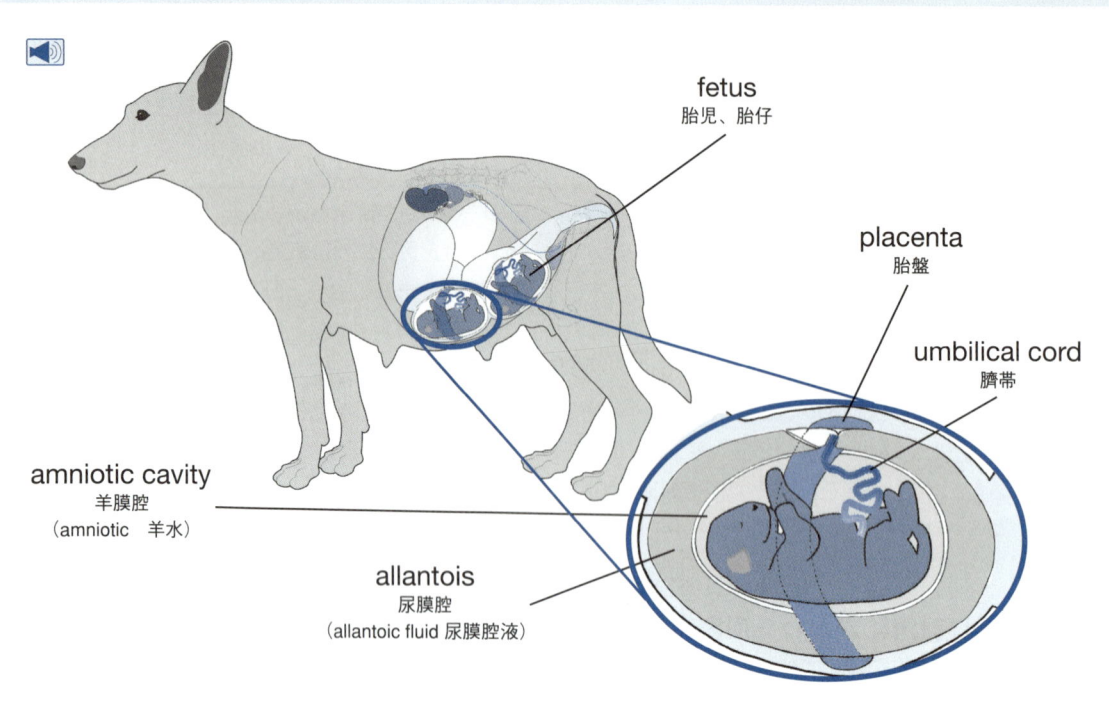

- fetus 胎児、胎仔
- placenta 胎盤
- umbilical cord 臍帯
- amniotic cavity 羊膜腔 (amniotic 羊水)
- allantois 尿膜腔 (allantoic fluid 尿膜腔液)

発情 Estrous*

anestrus	無発情
silent heat, silent estrus	鈍性発情
short period estrus	短発情
persistent estrus, continued estrus, prolonged estrus	持続性発情

* 口語では発情が来ていることを"in heat"という。

受精と排卵 Ovulation and Fertilization

ovulation		排卵
conception		受胎
fertilization 受精	fertility	受精能、受胎能
	artificial insemination（AI）	人工授精
	embryo transfer（ET）	受精卵移植、胚移植
	in vitro fertilization（IVF）	体外受精
	ovum pick up（OPU）	生体内卵子吸引
	embryonic stem cell	ES細胞、胚性幹細胞
rectal palpation		直腸検査（牛、馬、豚）
breeding		繁殖、交配
mating		交配
copulation / copulte		交尾／交尾する
repeat breeder*		リピート・ブリーダー

* 原因不明で、なかなか受胎しないメスのことを指す。発情は定期的に来ているにもかかわらず、原因不明で妊娠しないメスの場合に用いるため、卵巣疾患などの原因が分かれば、リピート・ブリーダーとは言わない。

妊娠 Pregnancy 🔊

implantation	着床
placenta	胎盤
embryo／fetus	胚／胎児
umbilical cord	臍帯
pseudopregnancy	偽妊娠
extrauterine pregnancy, ectopic pregnancy	子宮外妊娠

胎膜 Fetal Membrane 🔊

amnion／amniotic fluid	羊膜／羊水
allantois／allantoic fluid	尿膜／尿膜腔液
chorion	絨毛膜
chorioallantoic placenta	絨毛膜尿膜胎盤

分娩 Delivery 🔊

-tocia[*1]	eutocia	正常分娩[*2]
	dystocia	難産
abnormal parturition		異常分娩
labor pains		陣痛
uterine inertia		陣痛微弱、子宮無力症
hysterorrhexis		子宮破裂
uterine torsion		子宮捻転
endometritis		子宮内膜炎
cesarean section[*3]（C-section）		帝王切開
retained placenta		胎盤停滞
afterbirth		後産
premature birth		早産
stillbirth		死産
teratology, congenital anomalies, congenital defects		先天異常

[*1] -tocia は「分娩」を表す接尾辞。
[*2] eu は「正」を表す接頭語。
[*3] Cesarean と大文字で始めることもある。カエサル（Julius Caesar）が帝王切開で産まれたという伝承より、人名なら大文字、一般名詞ととらえるなら小文字でよい。

中絶 Abortion 🔊

castrate, neuter	去勢する[*1]
spay	避妊
ovariohysterectomy（OH）	卵巣子宮切除術[*2]
miscarriage	流産[*3]
abortion, artificial abortion	中絶、人工中絶[*4]

[*1] castrate は産業動物に、neuter はペットに使うことが多い。
[*2] 頻用されている避妊法である。
[*3] 期待に反した自然流産。
[*4] 意志をもって妊娠の継続をやめる行為。

Exercises 9

問1. 左右を正しく結びなさい。

1. ①精子 ・　　　　　・a) prostata, prostate
   ②精索 ・　　　　　・b) epididymis
   ③陰嚢 ・　　　　　・c) sperm
   ④前立腺 ・　　　　・d) spermatic cord
   ⑤精巣上体 ・　　　・e) scrotum

2. ①卵管 ・　　　　　・a) corpus luteum
   ②卵管采 ・　　　　・b) oocyte
   ③黄体 ・　　　　　・c) ovarian follicle
   ④卵胞 ・　　　　　・d) granular cell
   ⑤卵母細胞 ・　　　・e) fimbriae of uterine tube
   ⑥顆粒細胞 ・　　　・f) oviduct, uterine tube

問2. 次のホルモンの名称を日本語で書きなさい。

① LH（luteinizing hormone）：＿＿＿＿＿＿＿＿＿＿＿＿＿＿＿＿
② FSH（follicule-stimulating hormone）：＿＿＿＿＿＿＿＿＿＿＿＿
③ GnRH（gonadotropin-releasing hormone）：＿＿＿＿＿＿＿＿＿＿

問3. 左右を正しく結びなさい。

1. ①性成熟 ・　　　　・a) rectal palpation, rectal examination
   ②季節繁殖 ・　　　・b) puberty
   ③春機発動 ・　　　・c) sexual maturation
   ④直腸検査 ・　　　・d) seasonal breeding

2. ①精巣下降 ・　　　・a) semen
   ②停留精巣 ・　　　・b) frozen semen
   ③凍結精液 ・　　　・c) cryptorchidism
   ④精液 ・　　　　　・d) orchiocatabasis, discent of testis

3. ①人工授精 ・　　　・a) pseudopregnancy, false pregnancy
   ②偽妊娠 ・　　　　・b) artificial insemination (AI)
   ③（動物を）交尾させる ・　・c) embryo transfer (ET)
   ④受精卵 ・　　　　・d) mating
   ⑤受精卵移植（胚移植）・　・e) fertilized ovum

Exercises 9 の答え

問1．1．① c)　② d)　③ e)　④ a)　⑤ b)　　2．① f)　② e)　③ a)　④ c)　⑤ b)　⑥ d)
問2．①黄体形成ホルモン　②卵胞刺激ホルモン　③性腺刺激ホルモン放出ホルモン
問3．1．① c)　② d)　③ b)　④ a)　　2．① d)　② c)　③ b)　④ a)　　3．① b)　② a)　③ d)　④ e)　⑤ c)

Chapter 10

神経系および内分泌系とそれらの疾患
The Nervous and Endocrine Systems and their Disorders

1 神経系 The Nervous System

🔊

central nervous system (CNS) 中枢神経系	brain	脳
	spinal cord	脊髄
peripheral nervous system (PNS)　末梢神経系		

🔊

somatic nervous system 体性神経系	motor nerve　(=motoneuron)　*1	運動神経
	sensory nerve　(=sensory neuron)　*2	知覚神経
autonomic nervous system 自律神経系	sympathetic nerve	交感神経
	parasympathetic nerve	副交感神経

*1　afferent（求心性［輸入］神経線維）である。
*2　efferent（遠心性［輸出］神経線維）である

a) 神経 Nerve

語幹	英語名詞	英語形容詞	ラテン語名詞	ギリシャ語名詞
neur(o)-, neuri-	nerve(sg.), nervi(pl.)	nervous	nervus	neuron

神経にかかわる用語 Nerve Terminology 🔊

	neuron（=nerve cell）	神経細胞
	neuralgia, **neur**odynia	神経痛
	neuritis(sg.), **neur**itides(pl.)	神経炎
neur(o)-	**neur**ology	神経学
	neuroma	神経腫
	neurosis	ノイローゼ
	neuropathy	神経症、神経障害
	neurogenic shock	神経原性ショック

脳神経 Cranial Nerves 🔊

	I	olfactory nerve	嗅神経
	II	optic nerve	視神経
	III	oculomotor nerve	動眼神経
	IV	trochlear nerve	滑車神経
	V	trigeminal nerve	三叉神経
cranial nerves	VI	abducent nerve	外転神経
	VII	facial nerve	顔面神経
	VIII	vestibulocochlear nerve	内耳神経
	IX	glossopharyngeal nerve	舌咽神経
	X	vagus nerve	迷走神経
	XI	accessory nerve	副神経
	XII	hypoglossal nerve	舌下神経

b) 神経細胞、神経膠細胞、神経線維
Neuron, Glial Cell and Nerve Fiber

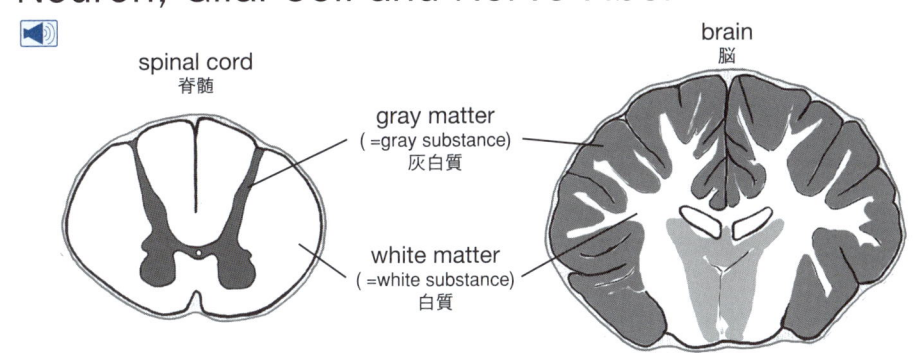

spinal cord
脊髄

brain
脳

gray matter
(=gray substance)
灰白質

white matter
(=white substance)
白質

脊髄では外側が白質、芯の部分が灰白質である。脳では逆に周辺部が灰白質、その内側が白質である。体積の大きい方が表面にくる。

白質と灰白質 White Matter and Gray Matter

white matter	白質*1
gray matter	灰白質*2

*1 主要構成要素は nerve fiber（神経線維）、とりわけ有髄線維（myelinated fiber）である。
*2 主要構成要素は nerve cell（神経細胞）である。

神経細胞、神経膠細胞 Nerve Cells and Gliocytes

	central nervous system （CNS）中枢神経系	peripheral nervous system （PNS）末梢神経系	和訳
nerve cell 神経細胞	nerve cell (=neuron) *	nerve cell (=neuron) ganglion cell	神経細胞（ニューロン） 神経節細胞
gliocyte (=glial cell) 神経膠細胞	astrocyte（=astroglia）	-	星状膠細胞
	oligodendrocyte (=oligodendroglia）	-	稀突起膠細胞
	microglia	-	小膠細胞
	-	Schwann cell	シュワン細胞

* ニューロン（neuron）は同じスペルのギリシャ語 " neuron " に由来する。もともとは、神経、腱などを表す言葉だったが、紀元前4世紀には神経に限局して用いられるようになった。現在は神経細胞に限局しているが、これは1891年にドイツ人解剖学者 Waldeyer が神経細胞をギリシャ語の neuron にちなんで名付けたことに由来する。

シナプス Synapse

neurotransmitter		神経伝達物質
neuromodulator		神経伝達調節因子
receptor 受容体	adrenergic receptor (=adrenoceptor)	アドレナリン受容体
	muscarinic receptor	ムスカリン受容体

c) 脳脊髄液と髄膜 Cerebrospinal Fluid and Meninges

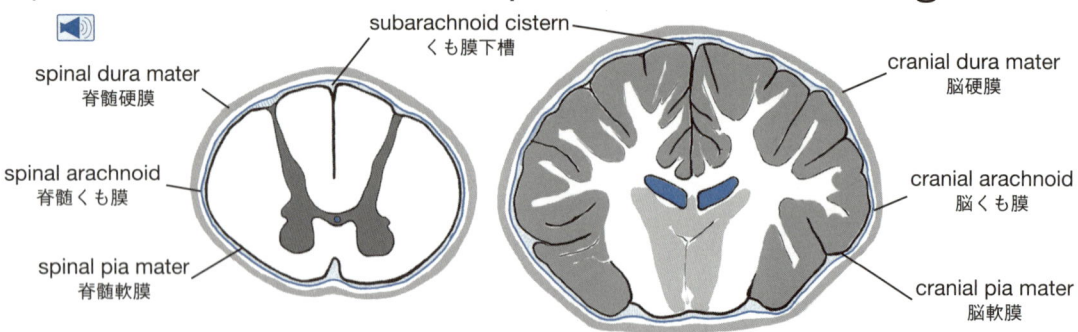

髄膜は meningis(sg.)、meninges(pl.)。ギリシャ語の meninx（膜）に由来する。

脳室 Ventricles

ventricles 脳室	lateral ventricle	側脳室
	third ventricle	第三脳室
	aqueduct of cerebrum	中脳水道
	fourth ventricle	第四脳室
spinal cord　脊髄	central canal	脊髄中心管
cerebrospinal fluid（CSF） 脳脊髄液	choroid plexus	脈絡叢
	arachnoid granulation	くも膜顆粒

d) 脳と脊髄 Brain and Spinal Cord

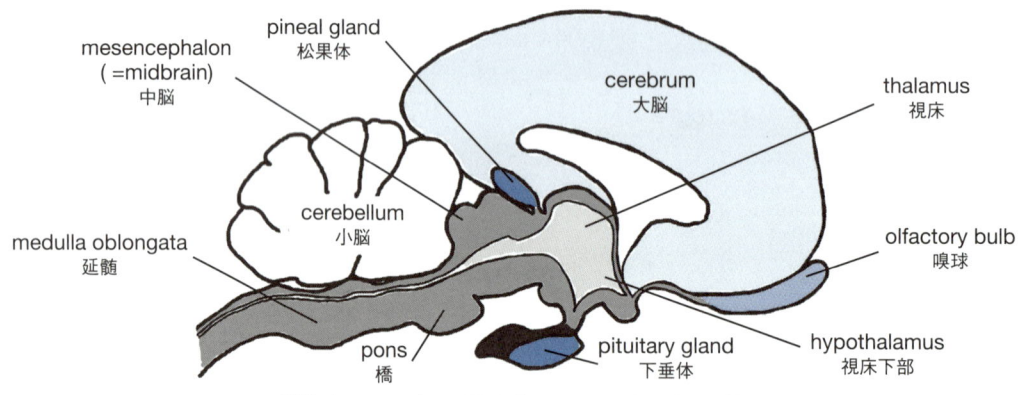

脳幹（brain stem）＝　間脳（diencephalon）＋中脳＋橋＋延髄。

脳 Brain

語幹	英語名詞	英語形容詞	ラテン語名詞	ギリシャ語名詞
encephal(o)-	brain	encephalic	encephalon	enkephalos

　脳（encephalon）は、en-（中）＋ kephale（頭）。頭の中にあるもの、という意味である。
　cerebrum はラテン語で「脳」という意味で、古くはサンスクリット語の ciras（頭尖）、ギリシャ語の keras（動物の角）に由来する。和訳は「大脳」である。この cerebrum に「小さい」という意味の "ll" を組み込んで作られた言葉が cerebellum（小脳）である。和訳では大脳、小脳と大小になっているが、もともと cerebrum という言葉には「大きい」という意味合いは無かった。

脳にかかわる用語 Cerebral Terminology 🔊

encephal(o)-	**encephalo**gram	脳造影図
	Japanese **encephal**itis	日本脳炎
cerebr-*¹	**cerebr**itis	脳炎
cerebral	cerebral infarction	脳梗塞
	cerebral thrombosis	脳血栓
cephal-*²	**cephal**itis (=**encephal**itis(sg.), **encephal**itides(pl.))	脳炎
	hydro**cephal**us	脳水腫、水頭症
その他	coma(sg.), comae (pl.)	昏睡
	epilepsy	てんかん
	narcolepsy	ナルコレプシー
	paralysis	麻痺
	spasm, convulsion	痙攣
	tetany	強直
	stroke	脳卒中

*1 「大脳」を表す語幹。
*2 「頭」を表す語幹。

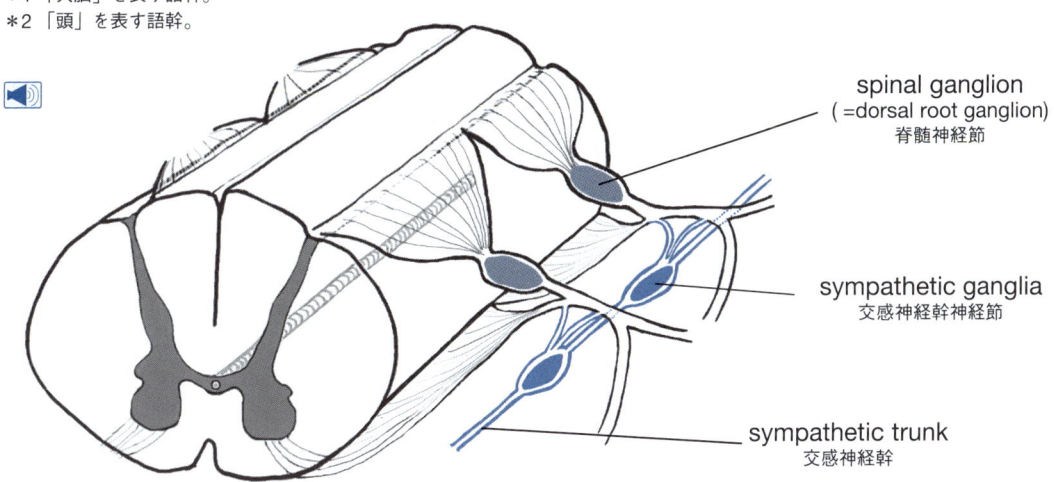

- spinal ganglion (=dorsal root ganglion) 脊髄神経節
- sympathetic ganglia 交感神経幹神経節
- sympathetic trunk 交感神経幹

脊髄 Spinal Cord

語幹	英語名詞	英語形容詞	ラテン語名詞	ギリシャ語名詞
myel-	spinal cord	myelic, myeloid, spinal	medulla spinalis	myelos

myel- には「髄」という意味しかない。骨髄も myel- という語幹をとるので注意。

脊髄にかかわる用語 Spinal Cord Terminology 🔊

myel-	**myel**algia	脊髄痛
	myelatrophy	脊髄萎縮
	myelitis	脊髄炎
その他	intervertebral disc herniation	椎間板ヘルニア
	spondylosis	脊椎症
	polyradiculoneuritis	多発性神経根神経炎

2 内分泌系 The Endocrine System

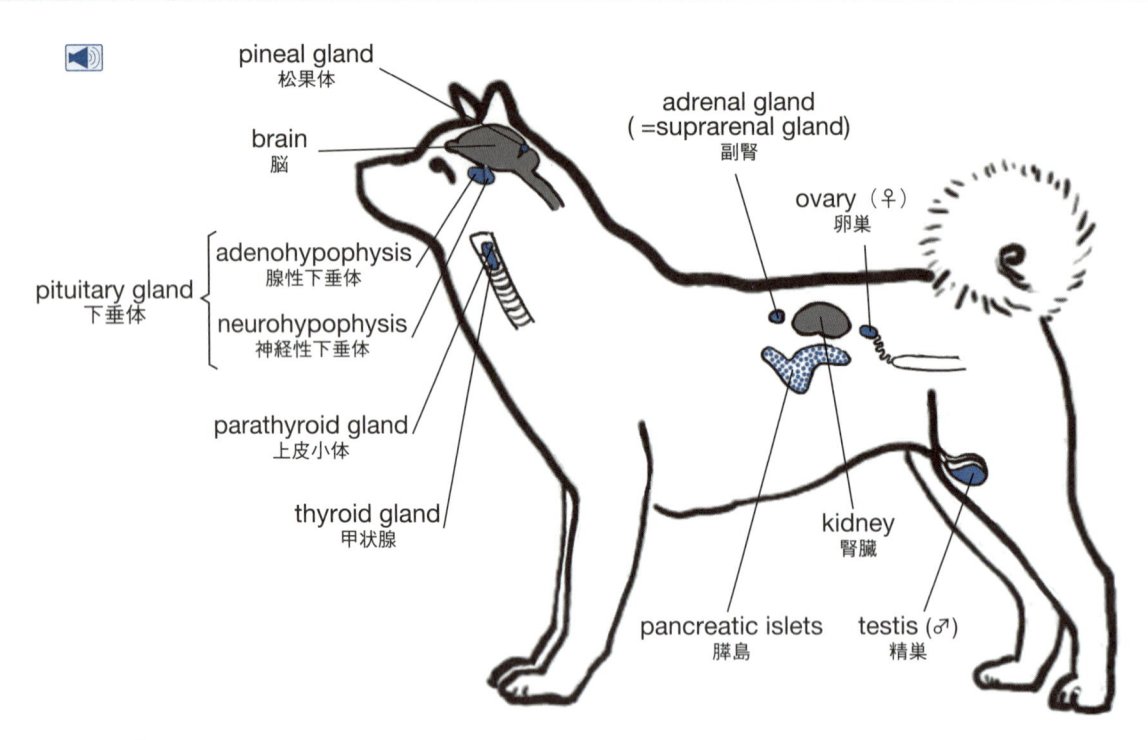

Diagram labels:
- pineal gland 松果体
- brain 脳
- adrenal gland (=suprarenal gland) 副腎
- ovary (♀) 卵巣
- pituitary gland 下垂体
  - adenohypophysis 腺性下垂体
  - neurohypophysis 神経性下垂体
- parathyroid gland 上皮小体
- thyroid gland 甲状腺
- kidney 腎臓
- pancreatic islets 膵島
- testis (♂) 精巣

下垂体 Pituitary Gland

語幹	英語名詞	英語形容詞	ラテン語名詞	ギリシャ語名詞
hypophys-, hypophyseo-	hypophysis, pituitary gland	hypophyseal, hypophysial, pituitary	hypophysis, glandula pituitaria	hypophysis

　下垂体は脳の直下に位置し、細い漏斗によって視床下部とつながっている。視床下部下垂体系は hypothalamo-hypophysial system と言う。

　hypophysis はギリシャ語起源で、hypo-（下）＋ physis（成長）である。別名 pituitary gland とも言う。pituita はラテン語で粘液、鼻汁を意味する。昔は脳が分泌した液が漏斗を通って下垂体に集まり、ここから嗅神経などを通じて鼻腔に運ばれた液が鼻汁だと考えられていたらしい。英語の pituitous は「粘液性の」という意味であって「下垂体の」ではないことに注意。

下垂体の疾患 Hypophyseal Disorders

hyperpituitarism, hyperfunction of pituitary	下垂体機能亢進症
pituitary gigantism, acromegaly	下垂体性巨大症、尖端巨大症
Cushing's disease[*1]	クッシング病
diabetes insipidus (DI)	尿崩症[*2]

＊1　クッシング症候群のうち、下垂体腺腫によるものを言う。
＊2　下垂体機能低下に伴う多飲、多尿。

内分泌器官とその疾患 Other Endcrine Organs and their Disorders 🔊

pineal gland 松果体	pinealoma	松果体腫
thyroid gland*1 甲状腺	thyroiditis	甲状腺炎
	hyperthyroidism	甲状腺機能亢進症*2
	hypothyroidism	甲状腺機能低下症
	goiter, struma	甲状腺腫
parathyroid gland*3 上皮小体	parastruma	上皮小体腫
	hyperparathyoridism, hyperparathyreodismus	上皮小体機能亢進症
	hypoparathyroidism	上皮小体機能低下症
adrenal gland *4 副腎	adrenal cortex	副腎皮質
	adrenal medulla	副腎髄質
	adrenalitis	副腎炎
	hyperadrenocorticism	副腎皮質機能亢進症
	Cushing's syndrome	クッシング症候群
endocrine pancreas*5 膵臓の内分泌部	diabetes mellitus	糖尿病
	insulinoma (insulin secreting islet cell tumor)	インスリノーマ (インスリン産生腫瘍)
gonad 性腺	Leydig cell tumor (interstitial cell tumor)	ライディッヒ細胞腫 (間質細胞腫)
	Sertoli cell tumor	セルトリー細胞腫
	hyperestrinism	エストロゲン過剰症

*1 甲状軟骨（thyroid cartilage）のそばにあることから、甲状腺と呼ばれるようになった。甲状軟骨は、腹側方から見た形が、古代ギリシャの兵士が持った盾"thyroes"に似ていることから sthyreoeidēs = thyr + oid（〜のような）と名付けられた。
*2 バセドウ病など。Basedow's goiter（バセドウ病甲状腺腫）：過剰のヨード摂取後に甲状腺機能亢進症となる甲状腺腫。
*3 parathyroid = para（近傍）+ thyroid（甲状）。甲状腺のそばにあることから。
*4 adrenal gland = ad-（近く）+ renal（腎臓の）。腎臓の近くの腺であることから。別名 suprarenal gland とも言う。 supra-（上）+ renal より、腎臓の上部の腺（人間だと上部になるが、動物は頭側）。
*5 pancreas = pan（すべて）+ creas（肉様）。インスリン（insulin）は、膵島（pancreatic islets）から分泌される。そこで、ラテン語の insula（島）+ -in（化学物質の接尾辞）より合成された言葉。

> **ホルモンの代表例 Hormones**
> - luteinizing hormone (LH)：黄体形成ホルモン
> - follicle stimulating hormone (FSH)：卵胞刺激ホルモン
> - thyroid stimulating hormone (TSH)：甲状腺刺激ホルモン
> - adrenocorticotropic hormone (ACTH)：副腎皮質刺激ホルモン
> - gonadotropin stimulating hormone：性腺刺激ホルモン
> - growth hormone：成長ホルモン
> - glucagon：グルカゴン
> - insulin：インスリン
> - progesterone：プロゲステロン（黄体ホルモン）
> - androgen：アンドロゲン（男性ホルモン）
> - estrogen：エストロゲン（卵胞ホルモン）

Exercises 10

問1．左右を正しく結びなさい。

1. ① ischemia　　　　　　　・　　　・ a) 脳貧血
   ② coma　　　　　　　　・　　　・ b) 虚血
   ③ depression　　　　　　・　　　・ c) 昏睡
   ④ cerebral anemia　　　　・　　　・ d) 脳充血
   ⑤ cerebral hyperemia　　　・　　　・ e) 脳出血
   ⑥ cerebral hemorrhage　　・　　　・ f) うつ、うつ病

2. ①脳卒中　　・　　　　・ a) stroke
   ②日射病　　・　　　　・ b) heatstroke
   ③熱射病　　・　　　　・ c) sunstroke

問2．それぞれの炎症名を英語で書きなさい。

① 甲状腺炎　＿＿＿＿＿＿＿＿＿＿＿＿＿
② 副腎炎　　＿＿＿＿＿＿＿＿＿＿＿＿＿
③ 下垂体炎　＿＿＿＿＿＿＿＿＿＿＿＿＿
④ 膵炎　　　＿＿＿＿＿＿＿＿＿＿＿＿＿

問3．それぞれの細胞の、腫瘍名を英語で書きなさい。

① 間質細胞腫（精巣）　＿＿＿＿＿＿＿＿＿＿＿
② インスリン産生細胞腫　＿＿＿＿＿＿＿＿＿＿
③ 下垂体細胞腫　＿＿＿＿＿＿＿＿＿＿＿＿＿

Exercises10の答え

問1．1. ① b)　② c)　③ f)　④ a)　⑤ d)　⑥ e)　2. ① a)　② c)　③ b)
問2．① thyroiditis　② adrenalitis　③ hypophysitis　④ pancreatitis
問3．① interstitial cell tumor　② insulinoma　③ pituicytoma

Chapter 11

感覚器系および外皮系とそれらの疾患

The Sensory and Integumentary Systems and their Disorders

1 視覚系 The Visual System

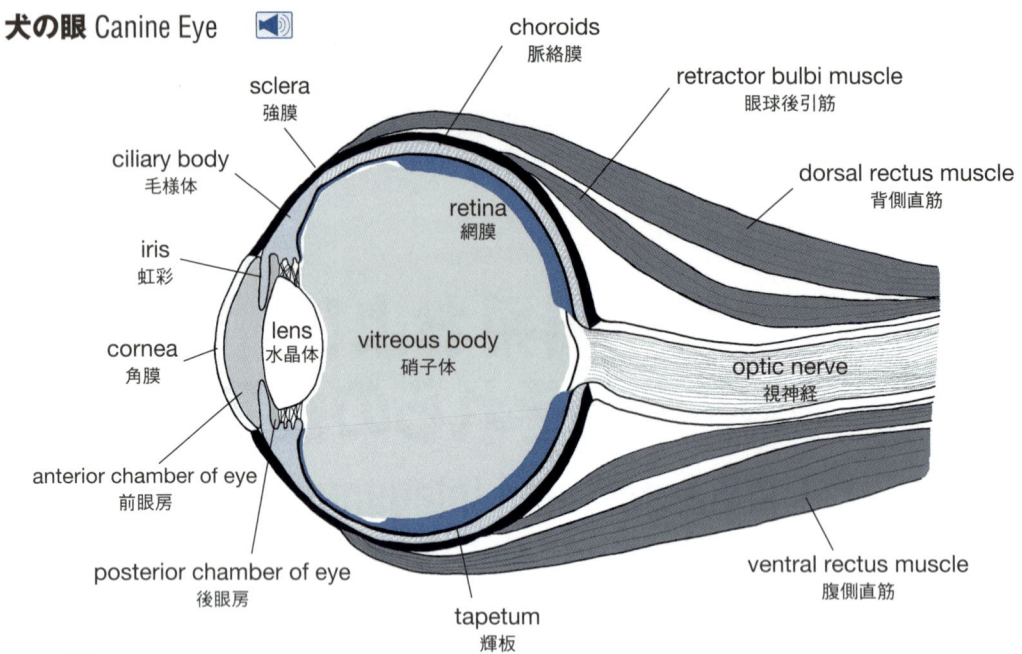

犬の眼 Canine Eye 🔊

(図:
- choroids 脈絡膜
- sclera 強膜
- ciliary body 毛様体
- iris 虹彩
- cornea 角膜
- lens 水晶体
- anterior chamber of eye 前眼房
- posterior chamber of eye 後眼房
- retina 網膜
- vitreous body 硝子体
- tapetum 輝板
- retractor bulbi muscle 眼球後引筋
- dorsal rectus muscle 背側直筋
- optic nerve 視神経
- ventral rectus muscle 腹側直筋
)

a) 眼 Eye

語幹	英語名詞	英語形容詞	ラテン語名詞	ギリシャ語名詞
ophthalm(o)-, optico-, opto-, oculo-	eye	ophthalmic, optic, ocular	oculus	ophthalmos, optikos*

　視覚という機能を表す英語は vision である。形容詞の visual は、ラテン語の visus（眺め）に由来する。 もともとは ophthalmic も optic も、「眼の」という意味であるが、脳神経では異なる神経を指すので注意が必要である。 optic nerve は第Ⅱ脳神経の視神経を、ophthalmic nerve は第Ⅴ脳神経三叉神経の一枝である眼神経を意味する。

　また、 ocular も「眼の」という意味を持ち、例えば ocular witness（目撃者）などのように使われる。解剖学用語では、「眼球（eye ball）の」というイメージを帯びる。第Ⅲ脳神経の動眼神経は oculomotor nerve である。

＊ optikos（眼の）はギリシャ語の属格（=英語の所有格）である。

眼にかかわる用語 Optic Terminology 🔊

ophthalm(o)-	**ophthalm**algia （=**oculo**dynia）	眼痛
	ophthalmia*	眼炎
optico-, opto-	**optico**kinetic nystagmus, **opto**kinetic nystagmus	視運動性眼振
	optometrist	検眼士
oculo-	**oculo**pathy （=**ophthalmo**pathy）	眼病

＊ 眼内炎（endoophthalmitis）、汎眼球炎（panophthalmitis）、眼周囲炎（periophthalmitis）という時は -itis が付く。単に眼炎という時は ophthalmitis という言い方もあるが、 ophthalmia の方が一般的。肺炎（pneumonia）と同様、 -ia で終わる。

148

b) 視覚器官 Other Visual Organs

角膜 Cornea

語幹	英語名詞	英語形容詞	ラテン語名詞	ギリシャ語名詞
kerat(o)-	cornea	corneal	cornea	-

網膜 Retina

語幹	英語名詞	英語形容詞	ラテン語名詞	ギリシャ語名詞
retin(o)-	retina	retinal	retina	-

強膜 Sclera

語幹	英語名詞	英語形容詞	ラテン語名詞	ギリシャ語名詞
scler(o)-	sclera	screral	sclera	skleros

虹彩 Iris

語幹	英語名詞	英語形容詞	ラテン語名詞	ギリシャ語名詞
irid(o)-	iris	iridal, iridial, iridian, iridic	iris	iris

副眼器 Accessory Ocular Organs 🔊

conjunctiva	結膜
eyelids	眼瞼、まぶた
eyelashes, cilia	まつ毛
semilunar fold of conjunctiva（=third eyelid, nictitating membrane）	結膜半月ヒダ（第三眼瞼、瞬膜）
lacrimal gland	涙腺

> **Column 虹の女神**
>
> "Iris"はギリシャ神話の虹の女神で、神々のメッセンジャーでもある。虹彩はこの女神の名前にちなんで付けられた美しい学名である。また、アイリス（あやめ）という花があるが、その名もこの女神に由来する。
>
> **カメラ**
>
> 眼の構造はしばしばカメラ（camera）にたとえられる。実際、眼の中には本当にカメラがあるのをご存知だろうか。角膜と水晶体の間のスペースを前眼房（anterior chamber of eye）、後眼房（posterior chamber of eye）というが、これらのラテン語学名は、それぞれ *Camera anterior bulbi*, *Camera posterior bulbi* なのである。英語の camera はアーチ型の天井をした小さな部屋（地下貯蔵室や、教会の地下納骨所）を意味するラテン語の camera、ギリシャ語の kamera、に由来する。眼房もアーチ型をしているし、暗くて小さい。ぴったりの名称と言えそうだ。

視覚器官にかかわる用語 Terminology of the other Visual Organs

kerat(o)-	**kerat**itis	角膜炎
retin(o)-	**retin**itis	網膜炎
	retinopathy	網膜症
retina	detachment of the retina	網膜剥離
	edema of the retina	網膜浮腫
scler(o)-	**scler**itis	強膜炎
irid(o)-	**irid**esis	虹彩炎
その他	progressive neurodeatrophia	進行性網膜萎縮
	conjunctivitis	結膜炎*1
	blindness, loss of one's eyesight	失明
	nyctalopia, night blindness	夜盲症
	cataract	白内障*2
	glaucoma	緑内障*3
	hemorrhage in the eyegrounds, fundal hemorrhage*4	眼底出血
	corneal ulcers	角膜潰瘍
	nebula*5	角膜混濁、角膜片雲
	sty	ものもらい（麦粒腫）
	entropion	眼瞼内反症*6
	ectropion	眼瞼外反症（赤眼）*7

*1 "pinkeye" と言うこともある。
*2 水晶体白濁。
*3 眼圧亢進による視神経乳頭圧迫。
*4 英語もラテン語も同じつづりの fundus は、底 (bottom) を意味する。眼底は fundus oculi (FO)、ocular fundus、胃底は fundus ventricul, fundus of stomach というようにである。ここから、「基本的な」という意味の fundamental という語も派生した。
*5 nebula はもともと「星雲」という美しい言葉だが、角膜がそのような状態になることも意味する。また他に尿の白濁を表す際にも用いられる。
*6 眼瞼と睫毛（しょうもう、まつ毛）が内側に変位し、直接角膜を刺激する病。
*7 結膜や角膜が異常露出する病。

Column 動物の感じる豊かな色の世界

しばしば動物はほとんど色盲だ、というような文章を目にする。確かに人は赤青緑の３原色を識別するのに対し、犬や猫や牛などは赤と青の２原色しか感知できず、しかも色を知覚する cone cell（錐状体細胞）の数が大変少ないので、色の識別能力は低い。これには哺乳類の祖先が、もともとは夜行性であったことに原因がありそうだ。一方、輝板（tapetum）はよく発達していて、薄明かりでも光をとらえやすくなっている。

しかし、鳥類や爬虫類はなんと４原色を識別できるのだそうだ。彼らの祖先だった恐竜や古代の爬虫類も、カラフルな世界を満喫していたのかもしれない。

2 平衡聴覚系
The Vestibular and Auditory Systems

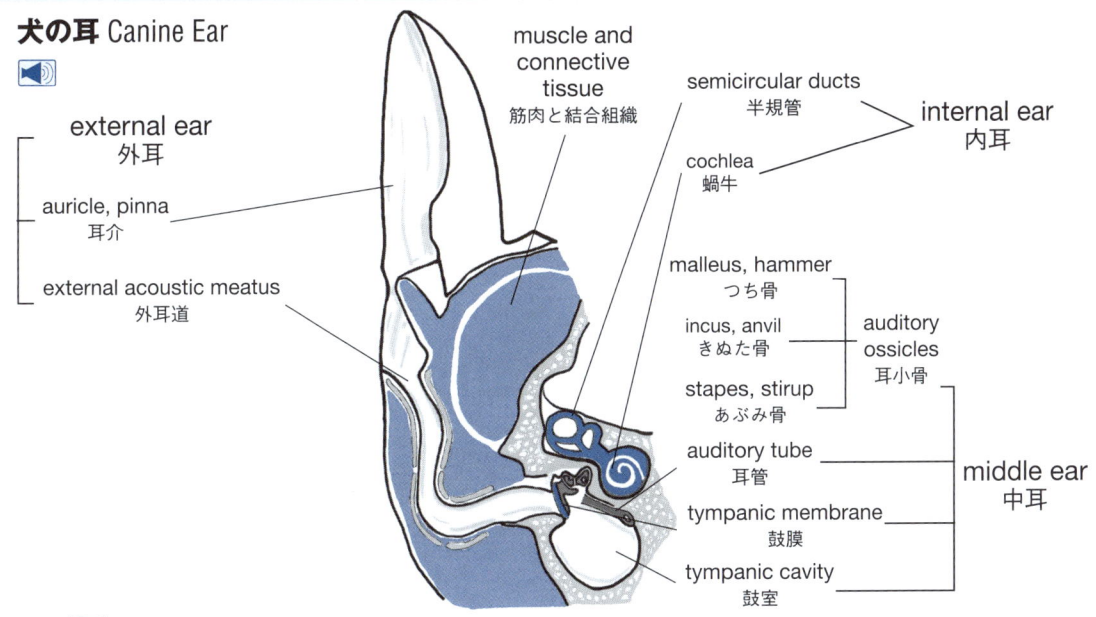

犬の耳 Canine Ear

- external ear 外耳
  - auricle, pinna 耳介
  - external acoustic meatus 外耳道
- muscle and connective tissue 筋肉と結合組織
- internal ear 内耳
  - semicircular ducts 半規管
  - cochlea 蝸牛
- middle ear 中耳
  - auditory ossicles 耳小骨
    - malleus, hammer つち骨
    - incus, anvil きぬた骨
    - stapes, stirup あぶみ骨
  - auditory tube 耳管
  - tympanic membrane 鼓膜
  - tympanic cavity 鼓室

a) 耳 Ear

語幹	英語名詞	英語形容詞	ラテン語名詞	ギリシャ語名詞
ot(o)-, auri-	ear	aural[*1], otic	auris	ōtos, ōtikos[*2]

聴覚という機能は audition、形容詞は auditory、acoustic である。

*1 aural の発音は、「口の」oral の発音と同じ。まぎらわしいので注意。
*2 ギリシャ語の属格（＝英語の所有格）。

耳にかかわる用語 Auricle Terminology

auri-	**auri**cle		耳介*
ot(o)-	**oto**acariasis		耳ダニ症
	otitis 耳炎	**ot**itis interna	内耳炎
		otitis media	中耳炎
		otitis externa	外耳炎
ear	inner ear （＝**auri**s interna）		内耳
	middle ear （＝**auri**s media）		中耳
	external ear （＝**auri**s externa）		外耳
	earache		耳痛

* 耳介を英語で pinna とも言うが、これはラテン語 pinna（羽）に由来。

b) 平衡聴覚器など Vestibulocochlear and other Organs

vestibulocochlear organ[*1]	平衡聴覚器
ceruminous gland	耳道腺
guttural pouch	喉嚢[*2]

*1 cochlea（蝸牛）はラテン語の cochlea, ギリシャ語の kokhlias（かたつむり）に由来。コルチ（Corti）器とも言う。蝸牛の組織学的詳細を初めて明らかにした、19世紀の解剖学者 Corti の名によるもの。
*2 耳管の一部が拡張し憩室となったもの。馬など奇蹄目に見る。

151

3 嗅覚系 The Olfactory System

a) 嗅覚 Olfaction

語幹	英語名詞	英語形容詞	ラテン語名詞	ギリシャ語名詞
olfacto-	olfaction	olfactory	olfactus	-

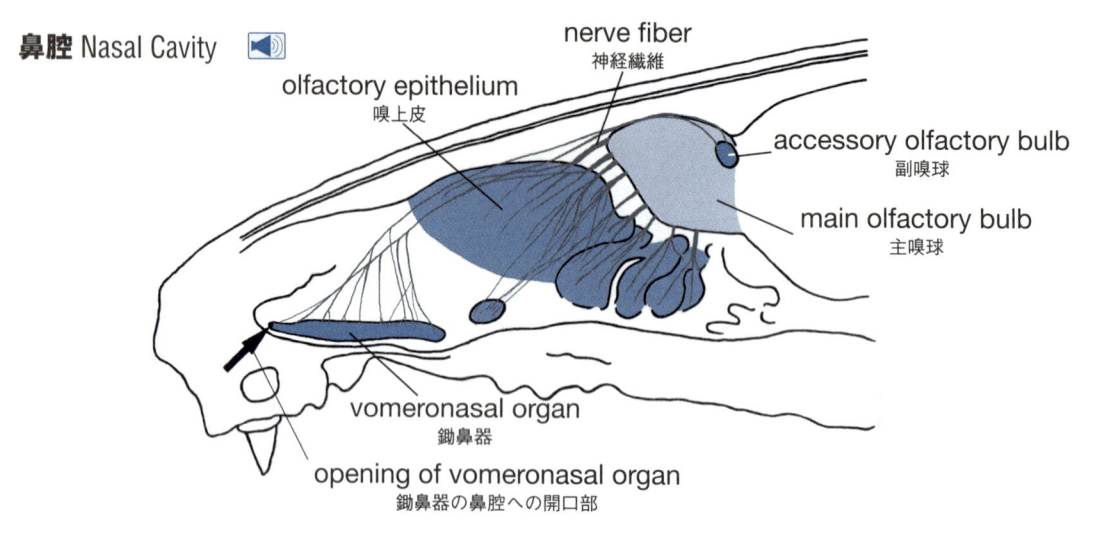

鼻腔 Nasal Cavity 🔊

- nerve fiber 神経繊維
- olfactory epithelium 嗅上皮
- accessory olfactory bulb 副嗅球
- main olfactory bulb 主嗅球
- vomeronasal organ 鋤鼻器
- opening of vomeronasal organ 鋤鼻器の鼻腔への開口部

b) 鼻 Nose

語幹	英語名詞	英語形容詞	ラテン語名詞	ギリシャ語名詞
rhin(o)-	nose	nasal	nasus	rhis

嗅覚および鼻にかかわる用語 Terminology of Olfaction and the Nose 🔊

olfacto-	**olfacto**metry	嗅覚検査
	olfactophobia	臭気恐怖症
rhin(o)-	**rhin**algia, **rhino**dynia	鼻痛
	rhinitis	鼻炎
	swine atrophic **rhin**itis	豚の委縮性鼻炎
その他	infectious coryza	伝染性コリーザ
	pheromone	フェロモン
	flehmen	フレーメン

Column 犀（サイ）

鼻に角のはえている動物である犀（rhinoceros）。rhino は鼻、ceros はギリシャ語の角（keras）である。角は皮膚の角質でできており、中に骨は無い。象に次ぐ大型哺乳類で体長4m、体重2.3トンにも達する巨体の持ち主である。その巨体が争う時は、この鼻の角をぶつけ合って戦うのである。

4 味覚系 The Gustatory System

a) 味覚 Gustation

語幹	英語名詞	英語形容詞	ラテン語名詞	ギリシャ語名詞
gust-	gustation	gustatory	gustatio	-

　gustatio はラテン語の gusto（味わうこと）に由来する。英語で味は taste であり、味覚は taste sensation と言う。味覚に関わるＧＴＰ結合蛋白質として最初に同定された alpha-gustducin（アルファガストドゥーシン）はこの gust- を語幹として名付けられた。

味蕾 Taste Bud 🔊

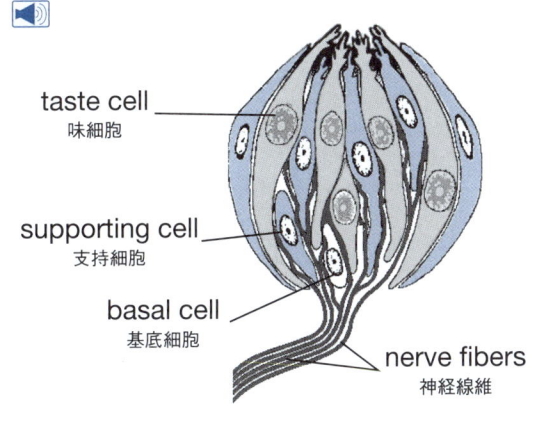

- taste cell 味細胞
- supporting cell 支持細胞
- basal cell 基底細胞
- nerve fibers 神経線維

b) 舌 Tongue

語幹	英語名詞	英語形容詞	ラテン語名詞	ギリシャ語名詞
lingu(o)-, gloss(o)-	tongue	lingual, glossal	lingua	glōssa

味覚および舌にかかわる用語 Terminology of Taste and the Tongue 🔊

lingual	lingual epithelium	舌上皮
lingu(o)-	**linguo**papilitis	舌乳頭炎
gloss(o)-	**gloss**algia, **glosso**dynia	舌痛
	glossitis	舌炎
taste	taste bud (=gustatory bud)	味蕾
	tastant	味物質

Column: Sushi の Umami

　日本語がそのまま英語になった言葉には、tsunami（津波）、bonsai（盆栽）、karaoke（カラオケ）、sushi（寿司）などがある。味覚の分野なら umami（旨味）であろう。この言葉と概念は、以前は懐疑的だった欧米の科学者たちには、もはやそのまま受け入れられるようになった。ただ、一般の人たちにはまだあまり知られていない。

5 外皮系 The Integumentary System

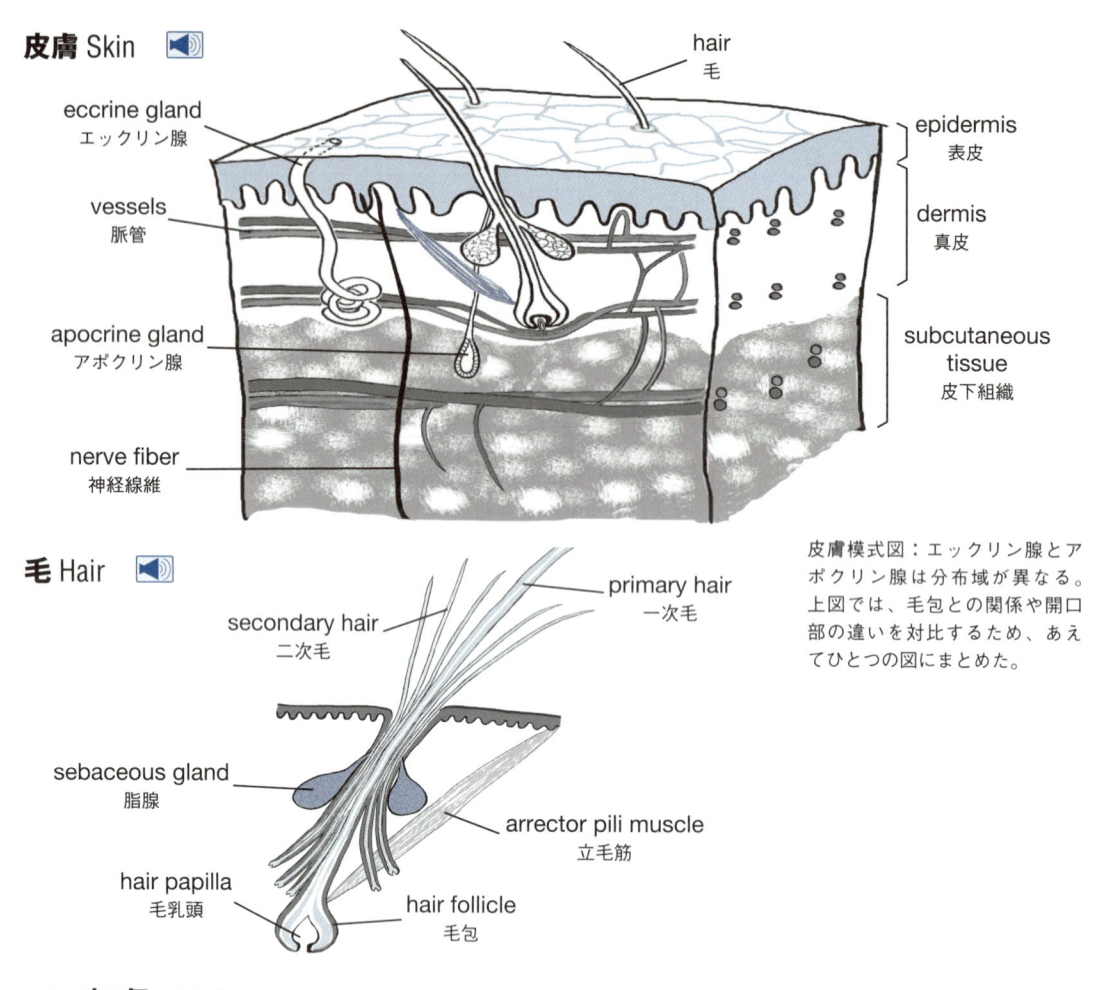

皮膚 Skin

- hair 毛
- eccrine gland エックリン腺
- vessels 脈管
- apocrine gland アポクリン腺
- nerve fiber 神経線維
- epidermis 表皮
- dermis 真皮
- subcutaneous tissue 皮下組織

毛 Hair

- primary hair 一次毛
- secondary hair 二次毛
- sebaceous gland 脂腺
- arrector pili muscle 立毛筋
- hair papilla 毛乳頭
- hair follicle 毛包

皮膚模式図：エックリン腺とアポクリン腺は分布域が異なる。上図では、毛包との関係や開口部の違いを対比するため、あえてひとつの図にまとめた。

a) 皮膚 Skin

語幹	英語名詞	英語形容詞	ラテン語名詞	ギリシャ語名詞
cuti-, derm-, derma-, dermat(o)-	skin, cutis	cutaneous, dermal	cutis	derma, dermatos*

真皮は dermis。表皮は真皮の「表（上）epi」にあるから、epidermis となる。"cuticle"（毛小皮、キューティクル）は cutis + cle（小さい）より。

＊ ギリシャ語の属格（＝英語の所有格）である。

角質層 Corneum

語幹	英語名詞	英語形容詞	ラテン語名詞	ギリシャ語名詞
kerat(o)-	corneum	corneal, keratic	cornu*	keras*

角質はケラチン（keratin）という蛋白質を主成分とする。角化は cornification、もしくは keratinization である。なお cornea は眼球の角膜で、これも角質のように硬い膜ということに由来する。このため、corneal には「角膜の」という意味もある。corn というと、「うおのめ」という意味もある。

＊ 「角」という意味を持つ。

表皮 Epidermis 🔊

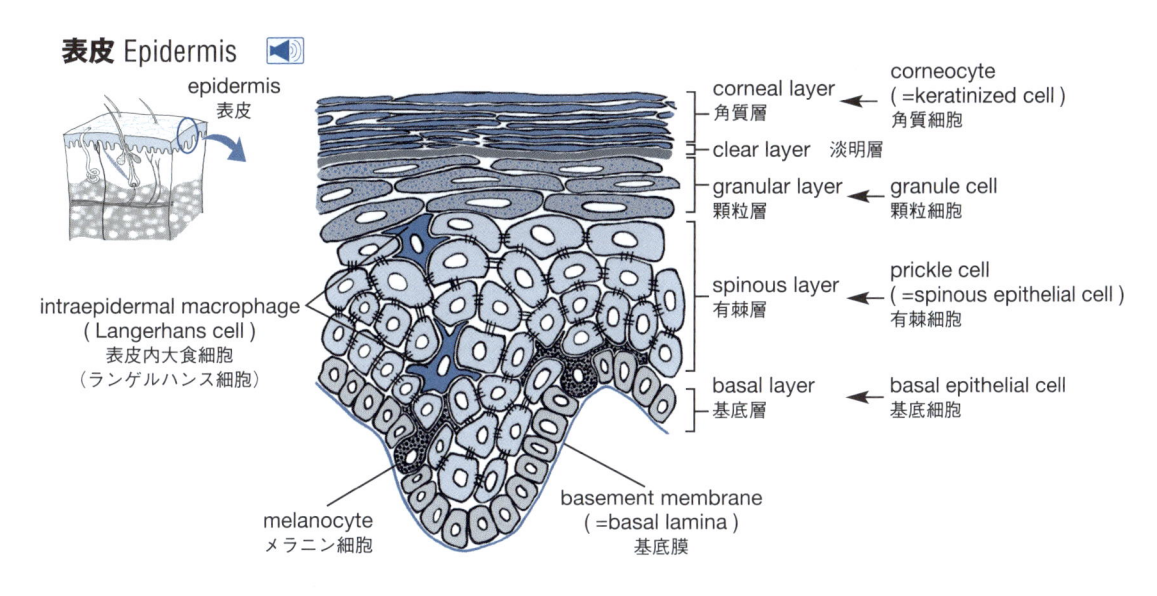

Diagram labels:
- epidermis 表皮
- corneal layer 角質層 — corneocyte (=keratinized cell) 角質細胞
- clear layer 淡明層
- granular layer 顆粒層 — granule cell 顆粒細胞
- spinous layer 有棘層 — prickle cell (=spinous epithelial cell) 有棘細胞
- basal layer 基底層 — basal epithelial cell 基底細胞
- intraepidermal macrophage (Langerhans cell) 表皮内大食細胞（ランゲルハンス細胞）
- melanocyte メラニン細胞
- basement membrane (=basal lamina) 基底膜

皮膚および角質層にかかわる用語 Terminology of the Epidermis and Keratin Layer 🔊

dermat(o)-	**dermat**algia, **dermato**dynia	皮膚痛
	dermatitis	皮膚炎
cuti-	**cuti**cle	毛小皮、クチクラ
	cutireaction	皮膚反応
kerat(o)-	**kerato**sis	角化症
	keratoma	角化腫

b) 爪、鉤爪、蹄 Nail, Claw and Hoof

特殊化した外皮 Specialized Integument 🔊

horn	角
pad	肉球
nasal planum	鼻鏡（犬など）

猫の鉤爪 Feline Claw 🔊

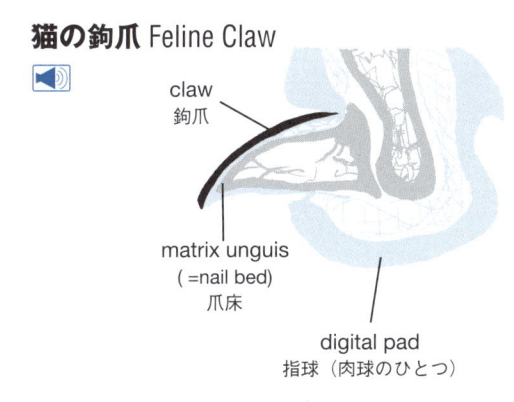

- claw 鉤爪
- matrix unguis (=nail bed) 爪床
- digital pad 指球（肉球のひとつ）

馬の蹄 Equine Hoof

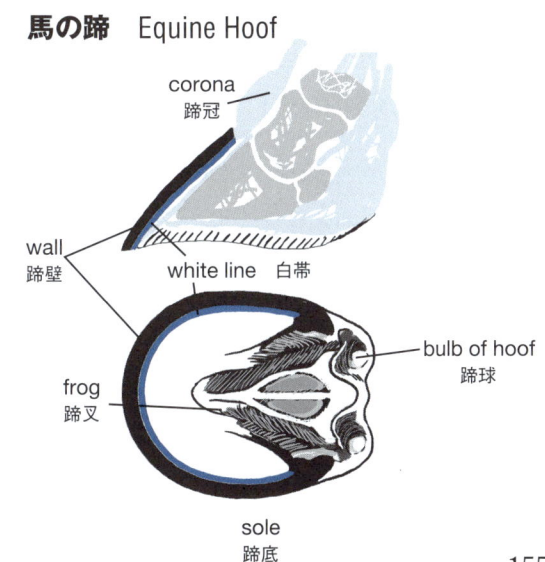

- corona 蹄冠
- wall 蹄壁
- white line 白帯
- bulb of hoof 蹄球
- frog 蹄叉
- sole 蹄底

爪、鉤爪、蹄 Nail, Claw and Hoof

nail		爪 （平爪）	ヒト、猿など
claw*1		鉤爪	食肉類（猫など）
hoof (sg.),*2 hooves (pl.) 蹄	coronary band	爪（蹄）冠	有蹄類（牛、馬など）
	wall	爪（蹄）底	
	sole	爪（蹄）壁	
	frog	蹄叉	
	white line	白帯*3	

*1 猫が claw で家具などを引っ掻いて傷つけないよう、declaw という手術を行うことがある。claw を除去（de-）する、という意味である。claw を切っただけではまた生えてくるので、指先の末節骨ごと切除する。アメリカではごく一般的に行われており、日本でも行われることがあるが、西欧では違法。

*2 代表的な蹄の疾病に laminitis（蹄葉炎）、canker（[馬] 蹄叉腐爛）などがある。

*3 white line を直訳すると「白線」と訳したくなる。しかし獣医学用語では「白帯」という訳語を当てる。これは英語ではなくラテン語の "zone alba" に従って訳したことになる。腹壁の "Linea alba（白線）" と同じ訳語になるのを避けるためにこうなった。

Exercises 11

問．左右を正しく結びなさい。

1. ① 失明 ・ ・a) diabetic retinopathy
   ② 眼底出血 ・ ・b) retinal detachment
   ③ 網膜剥離 ・ ・c) retinoplastoma
   ④ 糖尿病網膜症 ・ ・d) muscae volitantes
   ⑤ 飛蚊症 ・ ・e) blindness
   ⑥ 網膜芽細胞腫 ・ ・f) fundus hemorrhage

2. ① 難聴 ・ ・a) pollinosis
   ② 中耳炎 ・ ・b) otitis interna
   ③ 内耳炎 ・ ・c) sinusitis
   ④ 花粉症 ・ ・d) otitis media
   ⑤ 副鼻腔炎 ・ ・e) difficulty in hearing, deafness

3. ① にきび ・ ・a) acne
   ② 吹き出物 ・ ・b) dandruff
   ③ しわ ・ ・c) scar
   ④ かさぶた ・ ・d) chap, crack
   ⑤ 瘢痕 ・ ・e) blister
   ⑥ ふけ ・ ・f) wrinkle
   ⑦ ひび、あかぎれ ・ ・g) scab

Exercises11の答え

問1. 1. ① e) ② f) ③ b) ④ a) ⑤ d) ⑥ c)   2. ① e) ② d) ③ b) ④ a) ⑤ c)
3. ① a) ② e) ③ f) ④ g) ⑤ c) ⑥ b) ⑦ d)

Chapter 12

腫瘍学、病理学、寄生虫学
Oncology, Pathology and Parasitology

1 腫瘍学 Oncology

> 腫瘍を表す用語には、"onco-" を接頭辞に持つもの、"tumor" "cancer" で終わる2語以上のもの、"-carcinoma" "-sarcoma" などの接尾辞を持つものと、多様である。

腫瘍を表す用語　ただし、例外もあるので注意が必要。

onco-	腫瘍を表す接頭辞。「塊」を意味するギリシャ語名詞 "onkos" に由来する。
tumor	腫瘍。良性、悪性、上皮性、非上皮性のすべてを含む。neoplasm（新生物）とも言う。英語形容詞は tumorous。 「膨らむこと」を意味するラテン語名詞 "tumēre" に由来する。
cancer	癌。悪性で、上皮性の腫瘍を指す。英語形容詞は cancerous。「蟹」を意味するギリシャ語名詞 "cancer" に由来する。一説には、乳癌が蟹のように見えることからこの名が付いたという。ドイツ語で癌のことを Krebs というが、これも蟹の意味である。
-oma	上皮性、非上皮性（どちらも含む）の良性腫瘍を表す接尾辞。
-carcinoma	上皮性の悪性腫瘍（癌）を表す接尾辞。
-sarcoma	非上皮性の悪性腫瘍（肉腫）を表す接尾辞。

腫瘍にかかわる用語 Oncological Terminology

onco-	**onco**logy	腫瘍学
	oncogenesis	腫瘍形成＊
	oncogene	癌遺伝子
	oncovirus	腫瘍ウイルス

＊ carcinogenesis は発癌、すなわち悪性上皮性腫瘍の形成を意味する。

腫瘍学重要単語 Important Oncological Terms

metastasis(sg.), metastases(pl.)／metastasize	転移／転移する
anaplasia	退形成
dysplasia	異形成
atypia, atypism	異型性
parenchyma	実質
stroma	間質
differentiation／differentiated	分化／分化した

腫瘍の分類（代表例） Neoplasm

	benign 良性	malignant 悪性
epithelial tumor 上皮性腫瘍	-oma（腫）を含む語が多い papilloma：乳頭腫 adenoma：腺腫 epithelioma：上皮腫 sebaceous epithelioma 　：皮脂腺上皮腫 meibomian tumor 　：マイボーム腺腫 trichoepithelioma：毛包上皮腫 basal cell epithelioma 　：基底細胞（上皮）腫 polyp：ポリープ	-carcinoma（癌）を含む語が多い adenocarcinoma：腺癌 squamous cell carcinoma 　：扁平上皮癌 malignant seminoma 　：悪性精上皮腫 carcinoid：カルチノイド
non-epithelial tumor 非上皮性腫瘍	-oma（腫）を含む語が多い fibroma：線維腫 lipoma：脂肪腫 osteoma：骨腫 chondroma：軟骨腫 leiomyoma：平滑筋腫 myxoma：粘液腫 histiocytoma：組織球腫 angioma：血管腫 oligodendroglioma 　：稀突起膠細胞腫 meningioma：髄膜腫	-sarcoma（肉腫）を含む語が多い fibrosarcoma：線維肉腫 liposarcoma：脂肪肉腫 osteosarcoma：骨肉腫 condrosarcoma：軟骨肉腫 leiomyosarcoma：平滑筋肉腫 malignant histiocytoma 　：悪性組織球腫 angiosarcoma：血管肉腫 synovial sarcoma：滑膜肉腫 leukemia：白血病 multiple myeloma 　：多発性骨髄腫 malignant lymphoma 　：悪性リンパ腫 malignant melanoma 　：悪性黒色腫

ウイルス性腫瘍（代表例） Viral Tumors

		原因ウイルス	疾病名
RNA virus		feline leukemia virus（FeLV）	猫白血病・肉腫
		reticuloendotheliosis virus（REV）	細網内皮症
		avian leucosis virus, avian sarcoma virus（ALV, ASV）	鶏白血病・肉腫
		bovine leucosis virus（BLV）	牛白血病
DNA virus	papova virus パポーバウイルス	bovine papilloma virus（BPV）	牛の乳頭腫
		equine papilloma virus	馬の皮膚乳頭腫
		canine oral papilloma virus	犬の口腔乳頭腫
	herpes virus ヘルペスウイルス	Marek's disease	マレック病

2 病理学 Pathology

病理学重要単語 Important Pathological Terms

	dysbolism　代謝異常		
protein dysbolism 蛋白代謝異常	cloudy swelling		混濁腫脹
	hydropic degeneration		水腫変性
	vacuolar degeneration		空胞変性
	hyaline degeneration		硝子変性
	hyaline droplet degeneration		硝子滴変性
	fibrinoid degeneration		フィブリノイド（類線維素）変性
	amyloid degeneration		アミロイド変性
	keratin degeneration		角質変性
	mucinous degeneration		粘液変性
carbohydrate dysbolism 糖質代謝異常	diabetes		糖尿病
lipid and lipoid dysbolism 脂質・類脂質代謝異常	fatty degeneration, fatty change		脂肪変性（脂肪化）
	interstitial fatty infiltration		間質性脂肪浸潤
	lipidosis		脂質（蓄積）症
	xanthoma		黄色腫症
nucleic acid dysbolism 核酸代謝異常	gout		痛風
mineral dysbolism 無機物代謝異常	calcification 石灰沈着	hypocalcemia	低カルシウム血症
		hypercalcemia	高カルシウム血症
hyperpigmentation and dyspigmentation 色素過剰と色素沈着異常	hemoglobin		ヘモグロビン
	myoglobin		ミオグロビン
	hemosiderin		ヘモジデリン（鉄色素）
	bile pigment 胆色素	bilirubin	ビリルビン
		jaundice	黄疸
	lipofuscin		リポフスチン
	ceroid		セロイド
	melanin		メラニン
	cellular injury and death 細胞の障害と死		
atrophy 萎縮	–		–
differentiation disorder 細胞、組織分化異常	metaplasia		化生
	dysplasia		異形成
apoptosis アポトーシス[*1]	–		–

160

necrosis 壊死		coagulative (coagulation) necrosis	凝固壊死
		caseous necrosis	乾酪壊死
		gangrene	壊疽
		fat necrosis	脂肪壊死
		liquefactive necrosis (malacia)	液化壊死、融解壊死（軟化）
aging 老化、加齢		-	-
inclusion body formation 封入体形成		-	-
postmortem change 死後変化		autolysis	自己融解
		rigor mortis	死後硬直
		postmortem dots	死後凝血
		putrefaction	腐敗
progressive change 進行性変化			
hypertrophy			肥大
hyperplasia			増生、過形成
regeneration			再生
granulation tissues			肉芽組織
organization			器質化
foreign body			異物
elimination			排除
resorption			吸収
colliquation			融解
encapsulation			被包化
circulation disturbance 循環障害[*2]			
Inflammation 炎症			
effusion			滲出、滲出液
alterative inflammation, parenchymatous inflammation			変質性炎（実質性炎）
exudative inflammation			滲出性炎
hemorrhagic inflammation			出血性炎
purulent inflammation, suppurative inflammation			化膿性炎
fibrinous inflammation			線維素性炎
gangrenous inflammation			腐敗性炎（壊疽性炎）
proliferative inflammation			増殖性炎
granulomatous inflammation			肉芽腫性炎（特異性炎）
malformation 先天異常			

[*1] 細胞学は Chapter 4 （p.49〜）参照。
[*2] 血管、血液はChapter 6 （p.69〜）参照。

3 寄生虫学 Parasitology

> **Point** 寄生虫名の語尾に"-asis"か"-osis"が付くと、その寄生虫によって引き起こされる疾病名となる。日本語で寄生虫名の最後に「症」を付けると疾病名になることと同様である。

a) 内部寄生虫 Endoparasites

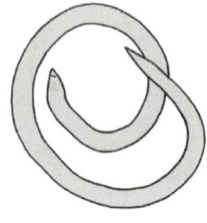

round worm　　　　　　tapeworm 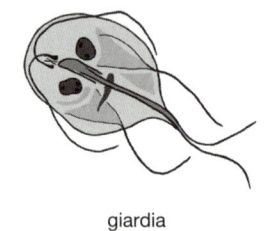　　　　　　giardia

🔊（青色は一般用語を表す）

Classification 分類	Parasite 寄生虫		Disease caused by the parasite それによって引き起こされる疾病	
protozoan, protozoon (sg.), protozoa, protozoans, protozoons(pl.)*1 原虫	babesia	バベシア	babesiasis (-osis)	バベシア症
	theileria	タイレリア	theileriasis	タイレリア症
	coccidium	コクシジウム類	coccidiosis (-asis)	コクシジウム症
	ameba, amoeba(旧名)	アメーバ	amebiasis, amoebiasis	アメーバ症
	trichomonas	トリコモナス	trichomoniasis	トリコモナス症
	giardia	ジアルジア	giardiasis	ジアルジア症
	toxoplasma	トキソプラズマ	toxoplasmosis	トキソプラズマ症
	piroplasma	ピロプラズマ	piroplasmosis	ピロプラズマ症
	trypanosoma	トリパノソーマ	trypanosomiasis	トリパノソーマ症
trematode, fluke 吸虫*2	*Fasciola hepatica* (=common liver fluke)	肝蛭	fascioliasis	肝蛭症
	liver fluke	肝吸虫	liver fluke disease	肝蛭症を含む肝臓の吸虫症の総称
	Paragonimus westermani	ウエステルマン肺吸虫	paragonimiasis westermani	ウエステルマン肺吸虫症
	lung fluke	肺吸虫	paragonimiasis (=lung fluke disease)	ウエステルマン肺吸虫を含む肺吸虫症の総称（肺ジストマ症ともいう）
	Schistosoma japonicum	日本住血吸虫	schistosomiasis japonicum	日本住血吸虫症
	blood fluke	住血吸虫	schistosomiasis (=blood fluke disease)	日本住血吸虫症を含む住血吸虫症の総称

cestode, tapeworm 条虫	diphyllobothrium	裂頭条虫	diphyllobothriasis	裂頭条虫症
	Taenia solium (=pork tapeworm, armed tapeworm)	有鉤条虫	Taenia solium taeniasis	有鉤条虫症
	tapeworm	条虫	taeniasis (=tapeworm disease)	条虫症
nematode 線虫	lungworm	肺虫	lungworm disease	肺虫症
	ancylostoma (=hookworm)	鉤虫	ancylostomiasis (=hookworm disease)	鉤虫症
	strongyloides	糞線虫類	strongyloidiasis	糞線虫症
	ascaris (=round worm)	回虫	ascariasis (=roundworm disease)	回虫症
	oxyurid (=pinworm)	蟯虫	oxyuriasis, (=pinworm disease)	蟯虫症
	filaria (=heartworm)	糸状虫	filariasis (=heartworm disease)	糸状虫症
	trichuris (=whipworm)	鞭虫	trichuriasis, (=whipworm disease)	鞭虫症
	anisakis	アニサキス	anisakiasis	アニサキス症
	trichinella (=trichina worm（旧名）)	旋毛虫 （トリヒナ）	trichinellosis, (=trichinosis)	旋毛虫症

＊1 protozoon = proto（最初の、原始の）＋ zoon（動物）より。
＊2 吸虫をジストマあるいは二口虫ともいう。ギリシャ語でdiは「2」、stomaは「口」。口吸盤と腹吸盤を両方とも口だと認識したことによる。

　表内の青色の英語と日本語をじっくり見比べてみよう。日本語の用語は、英語の直訳に驚くほど近いことがわかる。日本語の寄生虫の名称の多くが、おそらく英語由来なのだろう。例えば回虫を見てみよう。ascaris はギリシャ語の askaris（腸の虫）という意味であるが、ascaris のどこにも、「丸い」というイメージはない。だがこれが英語に訳されたとき、round worm となり、これが日本語に入り、回虫となったと思われる。
　では trichuris（鞭虫）はどうだろうか。これはギリシャ語の tricho-（髪の毛）と oura（尾）が合体した言葉である。「髪の毛のように細い尾を持つ」、が英語に訳されたとき、whip（ムチ）という言葉が選ばれ whipworm とされたのだろう。これを日本語では whip をそのまま鞭と直訳し、鞭虫となったのだ。

　Platyhelminthes（扁形動物）は flatworm とも呼ばれ、分類上の扁形動物門として使用される用語であり、条虫、吸虫の属する門を言う。Platy はギリシャ語の platys より「幅が広く、薄いもの」のことである。

　寄生虫の名称に -worm と付くものは、日常会話で使う一般名称であり、学術的な名称より直感的に覚えやすい。上の表には入れなかったが、-worm という名称を持つ寄生虫には他に eye worm（線虫の一種、地域によって検出に差異あり）や hair worm（ハリガネムシ）などがある。ハリガネムシ類（Gordiacea）は、家畜寄生性のものは無いとされるが、偶発例はある。

b) 外部寄生虫 Ectoparasites

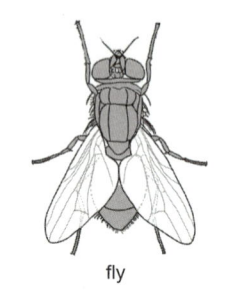

tick　　　　　　　　fly

🔊 （青色は一般用語を表す）

Classification	Prasite		Disease caused by the parasite	
Class arachnida 蛛形綱 Subclass acari* ダニ亜綱	ixodidae, ixodid tick, hard tick	マダニ	ixod**osis** ixodi**asis**	マダニ症
	sarcoptidae, mite	ヒゼンダニ	mange, scabies	疥癬
	demodicidae, demodex mite	ニキビダニ	demodic**osis**, demodicid**osis**	毛包虫症
	chigger mite	ツツガムシ	chigger dermatitis	ツツガムシ症
Class insecta 昆虫綱	fly (sg.), flies (pl.)	ハエ	myi**asis**	ハエウジ症
	blood sucking louse (sg.), lice (pl.)	シラミ	-	-
	biting louse (sg.), lice (pl.)	ハジラミ	-	-
	flea	ノミ	-	-

＊　ダニに当たる英語には tick と mite がある。tick は後気門亜目のダニで、通常大型で血を吸う。miteは無気門、前気門、中気門亜目のダニを指し、通常小型で吸血しない。

生活環 Life Cycle 🔊

intermediate host	中間宿主
final host, definitive host	終宿主
life cycle	生活環
budding	出芽
asexual reproduction	無性生殖
sexual reproduction	有性生殖
binary fission	二分裂
multiple fission	多数分裂
syngamy	融合
conjugation	接合

Exercises 12

問．左右を正しく結びなさい。

1. ①良性　　　　　・　　　・　a) malignant
   ②悪性　　　　　・　　　・　b) sarcoma
   ③新生物　　　　・　　　・　c) neoplasm
   ④転移　　　　　・　　　・　d) benign
   ⑤肉腫　　　　　・　　　・　e) metastasis

2. ①がん遺伝子　　・　　　・　a) oncology
   ②腫瘍学　　　　・　　　・　b) carcinostatic
   ③腫瘍マーカー　・　　　・　c) tumor marker
   ④抗がん剤　　　・　　　・　d) oncogene

3. ①萎縮　　　　　・　　　・　a) apoptosis
   ②変性　　　　　・　　　・　b) necrosis
   ③加齢　　　　　・　　　・　c) aging
   ④壊死　　　　　・　　　・　d) atrophy
   ⑤アポトーシス　・　　　・　e) degeneration

4. ①硝子変性　　　・　　　・　a) calcareous degeneration
   ②アミロイド変性・　　　・　b) hyaline degeneration
   ③石灰変性　　　・　　　・　c) amyloid degeneration

5. ①水腫、浮腫　　・　　　・　a) dehydration
   ②脱水　　　　　・　　　・　b) swelling
   ③膿疱　　　　　・　　　・　c) pustule
   ④腫脹　　　　　・　　　・　d) edema

6. ①終宿主　　　　・　　　・　a) ectoparasites
   ②中間宿主　　　・　　　・　b) endoparasites
   ③生活環　　　　・　　　・　c) intermediate host
   ④外部寄生虫　　・　　　・　d) life cycle
   ⑤内部寄生虫　　・　　　・　e) final host, definitive host

7. ①無性生殖　　　・　　　・　a) syngamy
   ②有性生殖　　　・　　　・　b) multiple fission
   ③二分裂　　　　・　　　・　c) sexual reproduction
   ④多数分裂　　　・　　　・　d) asexual reproduction
   ⑤出芽　　　　　・　　　・　e) binary fission
   ⑥融合　　　　　・　　　・　f) budding

Exercises12の答え

1. ① d)　② a)　③ c)　④ e)　⑤ b)　　2. ① d)　② a)　③ c)　④ b)　　3. ① d)　② e)　③ c)　④ b)　⑤ a)
4. ① b)　② c)　③ a)　　5. ① d)　② a)　③ c)　④ b)　　6. ① e)　② c)　③ d)　④ a)　⑤ b)
7. ① d)　② c)　③ e)　④ b)　⑤ f)　⑥ a)

付録 Appendices

Ⅰ. 専門分野と専門家 Departments and Specialists

　医学部では専門分野が細かく分かれており、例えば眼科の医者が産婦人科を兼ねることは、特殊な例を除いてまず無い。獣医学分野も、専門性を高める方向への動きがある。以下の表は、各科目とそこで働く専門医（専門家）の英語名をまとめたものである。

　一般的には「―学」は"-ology"、「―学の専門家」は"-ologist"となるが、例外もある。例えば外科は surgery、外科医は surgeon というようにである。physician には「内科医」という訳語が当てられている辞書もあるが、医師全般（外科医を除く）に用いられる。

Department 専門分野	Specialist 専門家	Japanese 和訳
anatomy	anatomist	解剖学（者）
anesthesiology	anesthesiologist	麻酔学（者）
anthropology	anthropologist	人類学（者）
bacteriology	bacteriologist	細菌学（者）
biochemistry	biochemist	生化学（者）
biotechnology	biotechnologist	生物工学（者）
cardiology	cardiologist	心臓科学（者）
dermatology	dermatologist	皮膚科学（者）
ecology	ecologist	生態学（者）
endocrinology	endocrinologist	内分泌学（者）
epidemiology	epidemiologist	伝染病学（者）
ethnology*1	ethnologist	民俗学（者）
ethology*1	ethologist	動物行動学（者）
forensic medicine（=legal medicine）	coroner（=medical attorney）	法医学（者）
genetics	geneticist	遺伝学（者）
gerontology	gerontologist	老年学（者）
gynecology*2	gynecologist	婦人科学（者）
hematology	hematologist	血液学（者）
histology	histologist	組織学（者）
hygiene	hygienist	衛生学（者）
ichthyology	ichthyologist	魚類学（者）
immunology	immunologist	免疫学（者）
internal medicine	internist	内科学（者）
microbiology	microbiologist	微生物学（者）
mycology	mycologist	真菌学（者）
molecular biology	molecular biologist	分子生物学（者）
morphology	morphologist	形態学（者）
neurology	neurologist	神経学（者）
obstetrics*2	obstetrician	産科学（者）

166

odontology（=dentistry）	odontologist（=dentist, dental surgeon）*3	歯学（者）
oncology	oncologist	腫瘍学（者）
ophthalmology	ophthalmologist（=an eye doctor）	眼科学（者）
ornithology	ornithologist	鳥類学（者）
orthopedics	orthopedist	整形外科学（者）
otorhinolaryngology*4	otorhinolaryngologist	耳鼻咽喉科学（者）
paleontology	paleontologist	古生物学（者）
parasitology	parasitologist	寄生虫学（者）
pathology	pathologist	病理学（者）
pediatrics	pediatrician	小児科学（者）
pharmacology	pharmacologist, pharmacist*5	薬理学（者）
physiology	physiologist	生理学（者）
plastic surgery	plastic surgeon	形成外科学（者）
proctology	proctologist	肛門病学（者）
psychiatry	psychiatrist	精神医学（者）
psychology	psychologist	心理学（者）
public health	public health specialist, public health administrator	公衆衛生学（者）
radiology	radiologist	放射線学（者）
serology	serologist	血清学（者）
surgery	surgeon	外科学（者）
teratology	teratologist	奇形学（者）
toxicology	toxicologist	毒性学（者）
urology	urologist	泌尿器科学（者）
virology	virologist	ウイルス学（者）
zoology	zoologist	動物学（者）

*1　ethology と ethnology とのスペル混同注意。
*2　ギリシャ語の gyne（女）に由来する。日本語では、しばしば産科と婦人科は産婦人科として存在しているが、アメリカなどでは産科学（obstetrics）と婦人科学（gynecology）は、全く独立した科として存在する。
*3　odontologist は学者、dentist や dental surgeon は歯科医（臨床家）のイメージが強い。
*4　otorhinolaryngology = otology（耳科学）+ rhinology（鼻科学）+ laryngology（喉頭科学）。
*5　pharmachologist は学者、pharmachist は薬剤師のイメージ。

Column 衛生の女神

衛生学（hygiene）という英語は、ギリシャ神話の健康、衛生の女神"ヒュギエイア（Hygieia, Hygeia）"（ギリシャ語の hugieinē）に由来する。医学や獣医学のシンボルとしてよく目にする、一本の杖に一匹の蛇がからみついているデザインがある。「アスクレピオス（Asclepius）の杖」と呼ばれ、太陽神アポロンの息子であるアスクレピオスのものである。アスクレピオスは医学をよくしたが、死者をもよみがえらせてしまったため、冥界の王"ハーデース（Hades = Pluto）"の怒りを買い、ゼウスの雷で殺されてしまう。死後その功績が再評価され、星座（へびつかい座）になった。ヒュギエイアはこのアスクレピオスの娘である。

Ⅱ. 獣医学臨床分野の頻出用語
Terms frequently used in Veterinary Clinical Fields

検査 Examinations

physical examination 身体的検査	inspection	視診
	auscultation	聴診
	percussion	打診
	palpation	触診
laboratory examination 臨床検査	blood test	血液検査
	urine examination	尿検査
	stool examination	糞便検査
	sputum examination	喀痰検査
	pathological examination	病理学的検査
	biopsy	バイオプシー、生検
	electrocardiogram（ECG）	心電図
	electroencephalogram（EEG）	脳波図
diagnostic imaging 画像解析	X-ray	X線
	computed tomography（CT）	コンピュータ連動断層撮影
	ultrasonography	超音波
	magnetic resonance imaging（MRI）	磁気共鳴画像法
	positron emission tomography（PET）	陽電子断層撮影法

注射 Injection

injection／inject	注射／注射する
intravenous（I.V.）*	静脈注射
intramuscular（I.M.）*	筋肉内注射
subcutaneous（S.C.）*	皮下注射
per os（P.O.）*	経口投与
vaccine／vaccination	ワクチン／ワクチン接種
transfusion	輸液
blood transfusion	輸血
intravenous drip（=I.V. drip）	点滴

＊ 大文字にして、省略のドットを付けるのが一般的だが、小文字でもよい。

全身症状 Constitutional symptom

recovery of activity	元気回復
hair loss	脱毛
poor appetite	食欲低下
weight loss	体重減少
obesity	肥満

薬 Medicine

antibiotics	抗生物質
anti-suppuration	化膿止め
antiflatulent	整腸剤
fever reducer	解熱剤
antidiarrhetic	下痢止め
laxative	下剤
painkiller	痛み止め
antipruritic	痒み止め
anesthetic	麻酔薬
sedative	鎮静薬
prescription diet	処方食

Ⅲ. ペットのリハビリ Rehabilitation

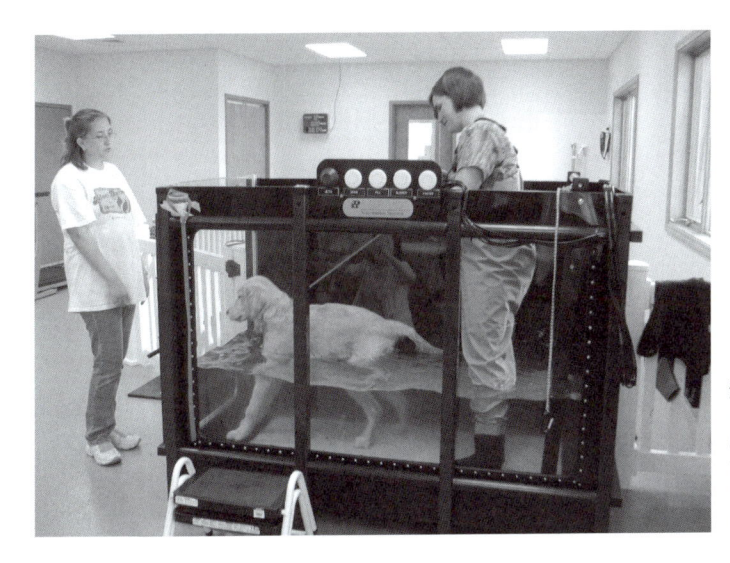

温水の中で犬を歩かせるリハビリ（下がベルトコンベアーになっていて、動く）を行っている、Dr. Stephanie Inoue (Indiana, USA)。運動療法と温熱療法を兼ねる。

リハビリテーションの目的 Purposes of Rehabilitation

- to improve the range of motion of joints：関節の可動性を高める。
- to speed up the healing process：手術後などの治癒過程を早める。
- to improve the function：機能を向上させる。
- to improve mental attitudes：ペットの精神面を活性化させる。
- to improve the quality of life：動物の生活の質を向上させる。

期待される結果 Expected Results

- reduced pain：痛みの軽減。
- reduced inflammation：炎症の軽減。
- reduced muscle and tissue atrophy：筋や組織の萎縮の軽減。

考慮すべき点 Points which should be considered

- age and physical conditions of the patient：患畜の年齢と生理学的状態。
- concurrent injuries：併発している障害はないか。
- owner compliance：ペットオーナーがどの程度協力的か。
- training as a therapist：獣医師自身の技量をよく見極めること。

方法 Methods 🔊

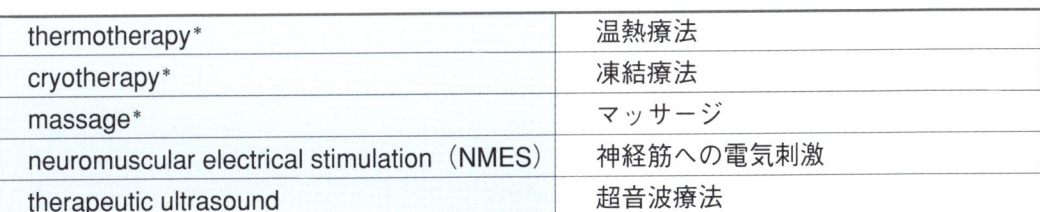

thermotherapy*	温熱療法
cryotherapy*	凍結療法
massage*	マッサージ
neuromuscular electrical stimulation（NMES）	神経筋への電気刺激
therapeutic ultrasound	超音波療法

* 徐々に受動的な可動領域"Passive Range of Motion（PROM）"を広げていきながら行う。

よくある症例 Frequently seen Cases 🔊

cranial cruciate ligament rupture	前十字靱帯断裂
bicipital tendonitis	上腕二頭筋腱炎
muscle strain	肉離れ、筋挫傷*
luxation（LX）	脱臼
subluxation	亜脱臼
elbow or hip dysplasia	肘関節や股関節の異形成
osteochondritis dessicans（OCDlesion）	離断性骨軟骨炎

＊ iliopsoas（腸腰筋）などでよく起こる。

動物にさせるエクササイズ Exercises for Animals

- assisted sitting, standing or walking：補助者が付いて座らせる、立たせる、歩かせる。
- physiorolls and balls：ボール遊びをさせる。
- cavaletti rails：ハードル（関節を曲げさせる、筋肉をなめらかに動かす）。
- tunnels：トンネルくぐり（ゲームの一種、匍匐前進をさせる）。
- wheelbarrowing／dancing：一輪車／ダンス。
- weights and therabands：セラバンド。
- land treadmills：トレッドミル。

> **Column 動物とアニメ**
>
> 　生きていれば、動物は息をする。ラテン語で息を anima と言い、これが転じて、「生命」という意味にもなった。animal（動物）はこの anima に由来する。
> 　anima に関連して、「生かす」というラテン語の動詞は animare で、ここから英語の animate（生命を吹き込む）が派生した。animate の名詞形が animation（アニメーション）である。生命を吹き込まれたもの、それがアニメなのである。

索引 Index

A

a set of false teeth	100
a tooth mark	100
abaxial	11, 19
abdomen	12, 15, 104
abdominal	15, 104
abdominal breathing	90
abdominal cavity	104
abdominal fluid	104
abdominal hernia	104
abdominal region	10
abdominal respiration	90
abdominal wall	104
abdominocentesis	38
abdominoscopy	104
abducent nerve	140
abduction	66
abductor muscle	66
ablactation	19
abnormal parturition	137
abomasitis	108
abomasum	107, 108
abortion	137
accessory nerve	140
accessory olfactory bulb	152
accessory pancreatic duct	119
achalasia	103
achillorrhaphy	38
acid	106
acidophilic leukocyte	80
acquired immune deficiency syndrome	86
acromegaly	144
actorrhea	35
acute pancreatitis	119
acute renal failure	123
acute rhinitis	91
acute ruminal dilatation	108
acute stage serum	80
acute (catarrhal) laryngitis	91
adduction	66
adductor muscle	66
adenocarcinoma	159
adenohypophysis	144
adenoma	159
adenosis	35
adherens junction	50
adhesive ileus	111
adiposome	50
adrenal cortex	145
adrenal gland	144
adrenal medulla	145
adrenalitis	145
adrenergic receptor	141
adrenoceptor	141
adventitia	76
adynamic ileus	111
afterbirth	137
agalactia	19
aggregated lymphatic nodule	82
aging	161
air brohchogram	94
albino	25
albumen	25
albumin	25
alcoholic cirrhosis	115
aliment	98
alimentary	98
alimentary canal	98
alimentation	98
alkaloid	32
allantoic fluid	137
allantois	136, 137
allergen	86
allergy	86
alloplasia	37
alpha cell	119
alterative inflammation	161
alveolar	95
alveolar bone	100
alveolar osteitis	101
alveolar pyorrhea	101
alveolar sac	95
alveoli	95
alveolitis	101
alveolus	95
ameba	162
amebiasis	162
amenorrhea	35
amniocele	36
amniocentesis	38
amnion	137
amniorrhexis	36
amniotic	136
amniotic cavity	136
amniotic fluid	137
amoeba	162
amoebiasis	162
amyloid degeneration	160
an artificial tooth	100
anal	45, 114
anal fistula	114
anaphylactic shock	86
anaphylaxis	86
anaplasia	37, 158
anastomosis	76
anatomist	166
anatomy	37, 166
ancylostoma	163
ancylostomiasis	163
anemia	19, 78
anesthesiologist	33, 166
anesthesiology	166
anesthetic	168
anestrus	136
aneurysm	76
an eye doctor	167
angina	75, 103
angina pectoris	75
angiogenesis	32, 75
angiography	75
angioma	75, 159
angiopathy	35
angioplasty	37
angiopoiesis	32
angiorrhaphy	38

angiosarcoma 75, 159	arachnoid 32	atrial premature complex 74
angitis 75	arachnoid granulation 142	atrial septal defect 72
ani 114	arachnophobia 34	atrio-ventricular 73
anisakiasis 163	arm 12	atrio-ventricular node 73
anisakis 163	arrector pili muscle 154	atrium 71
ankle 15	arrhythmia 73	atriums 71
ankylosing spondylosis 63	arterial 76	atrophia 33
ankylosis 63	arteriole 76	atrophy 19, 33, 160
antebrachial 15, 23, 58	arteriorrhexis 36	atypia 158
antebrachial region 10	arteriosclerosis 76	atypism 158
anterior 9, 23	artery 76	audiometer 31
anterior chamber of eye 148	arthralgia 46	audiometry 31
anthropologist 166	arthritis 43, 63	auditory ossicles 151
anthropology 166	arthrocele 36	auditory tube 151
anthropophobia 34	arthrodesis 63	aural 15, 151
anti-suppuration 168	arthrodynia 46	auricle 151
antibiotics 168	arthroplasty 63	auris externa 151
antibody 18, 85	arthropyema 63	auris interna 151
anticancer 18	arthrosis 63	auris media 151
antidiarrhetic 168	arthrostomy 38	auscultation 73, 168
antiflatulent 168	arthrotomy 63	autism 33
antigen 85	articular 43, 46, 62	autoimmune disease 86
antigen presenting cell 85	articular capsule 63	autolysis 35, 161
antigen-antibody reaction 85	articular cartilage 63	automated external defibrillator .. 73
antipruritic 168	articular fracture 63	autonomic nervous system 140
antiseptic 18	articular rheumatism 86	avian leucosis virus 159
anuria 125	articulation 62	avian sarcoma virus 159
anus .. 45, 98, 109, 112, 113, 114, 133	artificial abortion 137	AV node 73
anuses 114	artificial insemination 135, 136	axial 11
anvil 151	ascariasis 163	axillary lymph node 82
aorta 13, 71, 77, 79, 94	ascaris 163	axillary region 10
aortic incompetence 72	ascending colon 109	axis 11, 57
aortic insufficiency 72	ascites 104	azoospermia 135
aortic stenosis 72	asexual reproduction 164	azygos vein 77
aortic valve 71	aspiration pneumonia 95	
apical 52	astrocyte 141	**B**
apnea 90	astroglia 141	
apocrine gland 154	asystole 73	B cell 85
aponeurosis 65	atelectasis 95	babesia 162
aponeurositis 65	atlantooccipital joint 62	babesiasis（osis）............ 162
apoptosis 52, 160	atlas 57	baby 16
appendectomy 113	atony 66	back 10, 12, 15
appendical 112	atony of the forestomach 108	bacteremia 78
appendiceal 112	atony or impaction of abomasum .. 108	bacterial vaginosis 130
appendices 112	atopic asthma 86	bactericide 34
appendicitis 42, 113	atopic dermatitis 86	bacteriologist 166
appendicocele 36	atopy 86	bacteriology 166
appendicular 42	atria 71	basal 52
appendix 42, 112	atrial 71	basal cell 153
aqueduct of cerebrum 142	atrial fibrillation 73	basal cell epithelioma 159
		basal epithelial cell 155

basal lamina ············ 52, 155	blood-forming tissue ········· 81	calcification ············· 160
basal layer ················ 155	boar ······················ 16	calf ························ 16
basement membrane ···· 52, 155	body of stomach ·········· 105	calix ····················· 122
basocyte ··················· 80	body of vertebra ············ 57	calves ····················· 16
basophil ··················· 80	bone ············ 43, 46, 56, 57	calyx ···················· 122
basophile ·················· 80	bone ache ·················· 57	canine ················ 16, 102
basophilic leukocyte ········· 80	bone density ··············· 57	canine oral papilloma virus ···· 159
belly ······················ 15	bone fracture ············ 57, 59	canker sores ··············· 99
belly of muscle ············· 65	bone marrow ············ 57, 59	cannon ···················· 12
benign ···················· 159	bone trabecula ·············· 57	capillary ··················· 75
beta cell ·················· 119	bovine ····················· 16	capital ····················· 15
biceps ····················· 26	bovine leucosis virus ········ 159	caprine ···················· 16
bicipital tendonitis ··········· 170	bovine papilloma virus ······· 159	carbohydrate dysbolism ······ 160
bicipital tenosynovitis ········· 63	bowel ···················· 109	carcinogenesis ·············· 32
bicuspid valve ··············· 71	bowel movement ··········· 110	carcinoid ·················· 159
bile ················· 98, 116	bowel sounds ·············· 110	cardia ··········· 105, 106, 119
bile acid ·················· 116	Bowman's capsule ·········· 122	cardiac ··········· 14, 42, 45, 70
bile duct ·············· 116, 117	brachial ···················· 15	cardiac arrest ··············· 70
bile pigment ··············· 160	brachial region ·············· 10	cardiac gland ·············· 105
biliary ··············· 116, 117	brachiocephalic trunk ········· 79	cardiac hypertrophy ·········· 70
biliary duct ················ 117	bradycardia ················· 73	cardiac muscle ·············· 65
biliary sludge ··············· 118	brain ·· 14, 43, 46, 140, 141, 142, 144	cardiac myocyte ············· 70
bilirubin ··············· 116, 160	brain stem ················· 142	cardiac part ················ 106
binary fission ··············· 164	breast cancer ··············· 130	cardialgia ················ 45, 70
biochemist ················· 166	breech ····················· 15	cardio chalasia ·············· 106
biochemistry ··············· 166	breeding ··················· 136	cardiodynia ·············· 45, 70
biodynamics ················ 33	bridge of the nose ··········· 10	cardiologist ················ 166
biopsy ···················· 168	bronchi ···················· 94	cardiology ················· 166
biotechnologist ············· 166	bronchial ················ 42, 94	cardiomegaly ··············· 70
biotechnology ·············· 166	bronchiectasis ··············· 94	cardiomyopathy ·············· 70
bipolar ···················· 26	bronchitis ················ 42, 94	cardiopulmonary ············· 70
bisection ·················· 26	bronchopneumonia ··········· 94	cardiopulmonary resuscitation ·· 70
bladder ················ 46, 124	bronchoscope ··············· 30	cardiorrhexis ················ 36
（urinary）bladder ··········· 42	bronchoscopy ··············· 30	carditis ················· 42, 70
bladder stones ·········· 124, 125	bronchus ············ 42, 94, 95	carious teeth ··············· 100
bleeding ··················· 79	buccal ····················· 15	carnivore ··················· 58
blepharoptosis ··············· 36	budding ··················· 164	carpal ····················· 15
blindness ·················· 150	bulb of hoof ················ 155	carpal bones ·········· 12, 56, 58
bloat ··················· 99, 108	bulbitis ··················· 134	carpal joint ··············· 61, 62
blood ······················ 78	bulbourethral gland ·········· 133	carpal region ················ 10
blood capillary ··············· 75	bull ······················· 16	carpometacarpal joints ········ 61
blood coagulation ············ 78	bullae ····················· 95	cartilage ················· 43, 60
blood fluke ················ 162	bursa of Fabricius ············ 83	cartilage lacuna ·············· 60
blood formation ·············· 81	bursal ····················· 83	cartilage matrix ·············· 60
blood glucose level ··········· 79	buttocks ··················· 15	cartilaginoid ················ 60
blood smear ················ 78		cartilaginous ················ 43
blood sugar level ············ 79	# C	cartilaginous tissue ··········· 52
blood test ················· 168	caeca ····················· 112	caseous necrosis ············ 161
blood transfusion ····· 78, 86, 168	caecal ···················· 112	castrate ··················· 137
blood vessel ················ 75	calcaneal region ············· 10	cat ························ 16

173

cataract ⋯⋯⋯⋯⋯⋯⋯⋯ 150	cervical vertebrae ⋯⋯⋯⋯⋯ 57	chronic renal failure ⋯⋯⋯⋯ 123
catarrhal pneumonia ⋯⋯⋯⋯ 95	cervicitis ⋯⋯⋯⋯⋯⋯⋯⋯ 129	chronograph ⋯⋯⋯⋯⋯⋯⋯ 25
cats ⋯⋯⋯⋯⋯⋯⋯⋯⋯⋯ 16	cervix of uterus ⋯⋯⋯⋯ 127, 129	chylaemia ⋯⋯⋯⋯⋯⋯⋯⋯ 78
cattle ⋯⋯⋯⋯⋯⋯⋯⋯⋯⋯ 16	cesarean section ⋯⋯⋯⋯⋯ 137	chylothorax ⋯⋯⋯⋯⋯⋯⋯ 92
caudal ⋯⋯⋯⋯⋯⋯⋯⋯ 8, 9, 15	cestode ⋯⋯⋯⋯⋯⋯⋯⋯ 163	cilia ⋯⋯⋯⋯⋯⋯⋯⋯ 51, 149
caudal (posterior) vena cava ⋯ 71	cheek ⋯⋯⋯⋯⋯⋯⋯⋯⋯ 15	ciliary body ⋯⋯⋯⋯⋯⋯⋯ 148
caudal articular process ⋯⋯⋯ 57	chemotherapy ⋯⋯⋯⋯⋯⋯⋯ 39	circulation disturbance ⋯⋯⋯⋯ 161
caudal vena cava ⋯⋯⋯⋯ 13, 77	chest ⋯⋯⋯⋯⋯⋯⋯⋯ 10, 15	circulatory ⋯⋯⋯⋯⋯⋯⋯⋯ 78
caudal (coccygeal) vertebrae ⋯ 57	child ⋯⋯⋯⋯⋯⋯⋯⋯⋯⋯ 16	circulatory failure ⋯⋯⋯⋯⋯⋯ 76
cavity ⋯⋯⋯⋯⋯⋯⋯⋯⋯ 100	children ⋯⋯⋯⋯⋯⋯⋯⋯⋯ 16	circumflex ⋯⋯⋯⋯⋯⋯⋯⋯ 22
ceca ⋯⋯⋯⋯⋯⋯⋯⋯⋯⋯ 112	cholangiocarcinoma ⋯⋯⋯⋯ 118	circumvascular ⋯⋯⋯⋯⋯⋯ 22
cecal ⋯⋯⋯⋯⋯⋯⋯⋯⋯ 112	cholangiocellular carcinoma ⋯ 118	cirrhosis ⋯⋯⋯⋯⋯⋯⋯⋯ 115
cecitis ⋯⋯⋯⋯⋯⋯⋯⋯⋯ 113	cholangioma ⋯⋯⋯⋯⋯⋯⋯ 118	cisterna chyli ⋯⋯⋯⋯⋯⋯ 82
cecopexy ⋯⋯⋯⋯⋯⋯⋯⋯ 39	cholangiostomy ⋯⋯⋯⋯⋯⋯ 38	Class arachnida ⋯⋯⋯⋯⋯ 164
cecum ⋯⋯ 13, 98, 109, 112, 113	cholangitis ⋯⋯⋯⋯⋯⋯⋯ 118	Class insecta ⋯⋯⋯⋯⋯⋯ 164
celiac ⋯⋯⋯⋯⋯⋯⋯⋯⋯ 104	cholecyst ⋯⋯⋯⋯⋯⋯⋯⋯ 14	clavicle ⋯⋯⋯⋯⋯⋯⋯⋯⋯ 58
celiac artery ⋯⋯⋯⋯⋯⋯⋯ 104	cholecystalgia ⋯⋯⋯⋯⋯⋯ 118	claw ⋯⋯⋯⋯⋯⋯⋯ 155, 156
celiac lymph node ⋯⋯⋯⋯⋯ 82	cholecystectomy ⋯⋯⋯⋯ 116, 118	clear layer ⋯⋯⋯⋯⋯⋯⋯ 155
celiocentesis ⋯⋯⋯⋯⋯⋯ 104	cholecystic ⋯⋯⋯⋯⋯⋯⋯⋯ 42	clitoris ⋯⋯⋯⋯⋯⋯⋯⋯⋯ 129
celiorrhaphy ⋯⋯⋯⋯⋯⋯⋯ 104	cholecystis ⋯⋯⋯⋯⋯⋯⋯ 118	cloudy swelling ⋯⋯⋯⋯⋯ 160
celiotomy ⋯⋯⋯⋯⋯⋯⋯⋯ 104	cholecystitis ⋯⋯⋯⋯⋯⋯⋯ 42	cluneal ⋯⋯⋯⋯⋯⋯⋯⋯⋯ 15
cell ⋯⋯⋯⋯⋯⋯⋯⋯⋯⋯ 50	cholecystography ⋯⋯⋯⋯⋯ 118	cluneal region ⋯⋯⋯⋯ 10, 15
cell membrane ⋯⋯⋯⋯⋯⋯ 50	cholecystokinin ⋯⋯⋯⋯⋯ 116	coagulative (coagulation) necrosis ⋯ 161
cell wall ⋯⋯⋯⋯⋯⋯⋯⋯ 50	cholecystolith ⋯⋯⋯⋯⋯⋯ 118	coccidiosis (asis) ⋯⋯⋯⋯ 162
cellular ⋯⋯⋯⋯⋯⋯⋯⋯⋯ 50	cholecystolithiasis ⋯⋯⋯⋯⋯ 118	coccidium ⋯⋯⋯⋯⋯⋯⋯ 162
cellular immunity ⋯⋯⋯⋯⋯ 85	choledoch duct ⋯⋯⋯⋯⋯ 117	coccygeal vertebra ⋯⋯⋯⋯⋯ 12
cellular injury and death ⋯⋯⋯ 160	choledochitis ⋯⋯⋯⋯⋯⋯ 118	cochlea ⋯⋯⋯⋯⋯⋯⋯⋯ 151
cement ⋯⋯⋯⋯⋯⋯⋯⋯ 101	choledocholith ⋯⋯⋯⋯⋯⋯ 118	codon ⋯⋯⋯⋯⋯⋯⋯⋯⋯ 53
central ⋯⋯⋯⋯⋯⋯⋯⋯⋯ 11	choledocholithiasis ⋯⋯⋯⋯ 118	coefficient ⋯⋯⋯⋯⋯⋯⋯⋯ 20
central canal ⋯⋯⋯⋯⋯⋯ 142	cholelith ⋯⋯⋯⋯⋯⋯⋯⋯ 118	coffin bone ⋯⋯⋯⋯⋯⋯⋯ 58
central nervous system ⋯ 140, 141	cholelithiasis ⋯⋯⋯⋯⋯⋯ 118	coffin joint ⋯⋯⋯⋯⋯⋯⋯⋯ 62
centriole ⋯⋯⋯⋯⋯⋯⋯⋯ 51	cholestasia ⋯⋯⋯⋯⋯⋯⋯ 116	coffine bone ⋯⋯⋯⋯⋯⋯⋯ 62
cephalalgia ⋯⋯⋯⋯⋯⋯⋯⋯ 46	cholestasis ⋯⋯⋯⋯⋯⋯⋯ 116	cola ⋯⋯⋯⋯⋯⋯⋯⋯⋯ 112
cephalic region ⋯⋯⋯⋯⋯⋯ 10	cholorrhea ⋯⋯⋯⋯⋯⋯⋯⋯ 35	colic ⋯⋯⋯⋯⋯⋯⋯ 112, 113
cephalitis ⋯⋯⋯⋯⋯⋯⋯⋯ 143	chondral ⋯⋯⋯⋯⋯⋯⋯⋯ 60	colic lymph node ⋯⋯⋯⋯⋯ 82
cephalodynia ⋯⋯⋯⋯⋯⋯⋯ 46	chondralgia ⋯⋯⋯⋯⋯⋯⋯ 60	coliemia ⋯⋯⋯⋯⋯⋯⋯⋯ 113
cerebellum ⋯⋯⋯⋯⋯⋯⋯ 142	chondritis ⋯⋯⋯⋯⋯⋯ 43, 60	colitis ⋯⋯⋯⋯⋯⋯⋯⋯⋯ 113
cerebral ⋯⋯⋯⋯⋯⋯⋯⋯⋯ 14	chondroblast ⋯⋯⋯⋯⋯⋯⋯ 60	collagen diseases ⋯⋯⋯⋯⋯ 86
cerebral hemisphere ⋯⋯⋯⋯ 20	chondrocyte ⋯⋯⋯⋯⋯⋯⋯ 60	collagen fiber ⋯⋯⋯⋯⋯⋯ 52
cerebral infarction ⋯⋯⋯⋯⋯ 143	chondrodynia ⋯⋯⋯⋯⋯⋯⋯ 60	collagen-vascular diseases ⋯⋯ 86
cerebral thrombosis ⋯⋯⋯⋯ 143	chondroid ⋯⋯⋯⋯⋯⋯⋯⋯ 60	collagenosis ⋯⋯⋯⋯⋯⋯⋯ 86
cerebritis ⋯⋯⋯⋯⋯⋯⋯⋯ 143	chondroma ⋯⋯⋯⋯⋯⋯⋯ 159	collecting duct ⋯⋯⋯⋯⋯⋯ 122
cerebrospinal fluid ⋯⋯⋯⋯ 142	chorditis ⋯⋯⋯⋯⋯⋯⋯⋯ 91	colliquation ⋯⋯⋯⋯⋯⋯⋯ 161
cerebrum ⋯⋯⋯⋯⋯⋯ 14, 142	chorioallantoic placenta ⋯⋯⋯ 137	colocalization ⋯⋯⋯⋯⋯⋯ 20
ceroid ⋯⋯⋯⋯⋯⋯⋯⋯⋯ 160	chorion ⋯⋯⋯⋯⋯⋯⋯⋯ 137	colon ⋯⋯⋯ 13, 98, 112, 113, 119
ceruminous gland ⋯⋯⋯⋯⋯ 151	choroid plexus ⋯⋯⋯⋯⋯⋯ 142	colonic ⋯⋯⋯⋯⋯⋯⋯⋯ 112
cervical ⋯⋯⋯⋯⋯⋯⋯⋯⋯ 15	choroids ⋯⋯⋯⋯⋯⋯⋯⋯ 148	colonic diverticula ⋯⋯⋯⋯⋯ 113
cervical canal ⋯⋯⋯⋯⋯⋯ 129	chronic atrophic gastritis ⋯⋯⋯ 105	colons ⋯⋯⋯⋯⋯⋯⋯⋯ 112
cervical region ⋯⋯⋯⋯⋯⋯ 10	chronic laryngitis ⋯⋯⋯⋯⋯ 91	colospasm ⋯⋯⋯⋯⋯⋯⋯ 113
cervical stenosis ⋯⋯⋯⋯⋯ 129	chronic pancreatitis ⋯⋯⋯⋯ 119	colostomy ⋯⋯⋯⋯⋯⋯⋯⋯ 38

colostrum ·················· 131	corona ·················· 155	cutireaction ··············· 155
colpalgia ················ 46, 130	coronary artery ············· 77	cutis ······················ 154
colpatresia ················· 130	coronary band ············· 156	Cyanobacteria ·············· 25
colpectomy ················ 130	coronary thrombosis ········· 72	cyanosis ········· 25, 35, 76, 78
colpitis ················· 43, 130	coroner ···················· 166	cystalgia ················ 46, 124
colpodynia ··············· 46, 130	coronet ···················· 12	cysterna chili ················ 79
colposcope ·············· 30, 130	corpus luteum ············· 128	cystic ············ 14, 42, 46, 124
colposcopy ················· 30	cortex ····················· 122	cystic duct ················· 117
colt ························· 16	costal ······················ 15	cystic gall duct ············· 117
coma ···················· 143	costalgia ··················· 45	cystic kidney ··············· 123
comae ···················· 143	counterconditioning ·········· 18	cystitis ·················· 42, 124
combined immunodeficiency ·· 86	cow ······················· 16	cystodynia ·················· 46
commissure ················ 20	cows ······················ 16	cystogram ················· 124
common bile duct ······ 116, 117	coxae ······················ 46	cystolith ················ 124, 125
common carotid artery ········ 77	coxal ·················· 15, 46	cystoptosis ·················· 36
common hepatic duct ···· 115, 117	coxal bone ······· 12, 13, 56, 58	cytokine ···················· 85
complement ················ 85	coxalgia ···················· 46	cytometer ··················· 31
complement fixation ·········· 85	coxodynia ··················· 46	cytometry ··················· 31
component ················· 20	coxofemoral luxation ·········· 59	cytoplasm ··················· 50
conception ················ 136	cramp ····················· 66	cytoplasmic ················· 50
condrosarcoma ············· 159	cranial ··················· 8, 9	cytoskeleton ················· 51
congenital ·················· 20	cranial (anterior) vena cava ···· 71	cytosol ····················· 50
congenital anomalies ········ 137	cranial arachnoid ············ 142	
congenital defects ·········· 137	cranial articular process ······· 57	**D**
congestion ··············· 76, 78	cranial bone ················ 12	death ······················ 73
congestion of the liver ········ 115	cranial cavity ··············· 90	debarking ··················· 91
congestive heart failure ········ 70	cranial cruciate ligament rupture ·· 63, 170	decapitate ··················· 22
conjugation ················ 164	cranial dura mater ··········· 142	decayed teeth ··············· 100
conjunctiva ············· 43, 149	cranial pia mater ············ 142	deciduous teeth ············· 102
conjunctival ················· 43	cranial vena cava ·········· 13, 77	deferent duct ·········· 132, 133
conjunctivitis ············ 43, 150	cranium ···················· 56	deferential ················· 133
connective tissue ·········· 52, 76	cremaster muscle ············ 132	deferentitis ················· 134
constipation ················ 114	cross ······················ 9	defibrillation ················· 22
continued estrus ············ 136	croup ····················· 12	definitive host ··············· 164
continuous murmur ··········· 73	croupous laryngitis ··········· 91	degeneration ················ 52
contraception ················ 18	croupous pneumonia ·········· 95	degenerative muscular disease ·· 66
contraction ················· 66	crown of tooth ············· 100	delta cell ··················· 119
contralateral ················· 18	crural ······················ 15	demilune ···················· 20
convalescent stage serum ····· 80	crural region ················ 10	demodicidae ················ 164
convulsion ··············· 66, 143	cruris ······················ 58	demodicosis ················ 164
copulation ················· 136	cryotherapy ············· 39, 169	dendritic cell ················· 85
copulte ···················· 136	cryptorchism ··············· 132	densitometer ················· 31
cor ························ 70	computed tomography (CT) ·· 168	densitometry ················· 31
cor pulmonale ··············· 70	cubital ····················· 15	dental ··············· 14, 45, 100
cornea ················ 43, 148, 149	cubital region ··············· 10	dental caries ··············· 100
corneal ············· 43, 149, 154	Cushing's disease ··········· 144	dental pulp ·············· 42, 100
corneal layer ··············· 155	Cushing's syndrome ·········· 145	dentalgia ···················· 45
corneal ulcers ·············· 150	cutaneous ··············· 43, 154	dentine ···················· 101
corneocyte ················· 155	cutaneous muscle ············ 65	dental surgeon ··············· 167
corneum ··················· 154	cuticle ····················· 155	dentist ···················· 167

175

dentistry 100, 167	dipsomania 34	ecologist 166
dentures 100	discitis 63	ecology 166
deoxyribonucleic acid (DNA) .. 53	disinfect 22	ectopic pregnancy 129, 137
depression 66	dislocation 59	ectropion 150
depressor muscle 66	displaced abomasum 108	edema of the retina 150
dermal 43, 46, 154	dissect 22	electroencephalogram (EEG) 168
dermatalgia 46, 155	distal 11	effusion 161
dermatitis 43, 155	distal interphalangeal joints of manus .. 61	egomania 34
dermatodynia 46, 155	distal phalanx 58, 62	elastic cartilage 60
dermatologist 166	distal tubule 122	elastic fiber 52
dermatology 166	distoma 99	elbow 15
dermatomyositis 66	Distoma hepaticum 115	elbow joint 61, 62
dermis 154	disturbance of eructation relex .. 99	elbow or hip dysplasia 170
descending colon 109	disturbance of mastication and deglutition 99	electrocardiogram 31, 70
desmosome 50	disturbance of rumination 99	electrocardiograph 31
detachment of the retina 150	diverticula of colon 113	electroencephalogram 31
detoxication 22	DNA virus 159	electroencephalograph 31
diabetes 160	dog 16	elimination 161
diabetes insipidus 144	dog heartworm 70	embolic nephritis 123
diabetes mellitus 119, 145	dogs 16	embolism 76
diagnostic imaging 168	dorsal 8, 9, 11, 15	embryo 137
diamine 26	dorsal rectus muscle 148	embryo transfer 136
diaphragm 13, 22, 93	dorsal region 10	embryonic stem cell 136
diaphragmatic 93	dorsal root ganglion 143	empathy 22
diaphragmatic hernia 93	dorsal view 94	empyema 44, 91, 92
diaphragmatic view 94	drawing close 66	enamel 101
diaphysis 59	duodena 110	encapsulation 161
diarrhea 22, 35, 114	duodenal 110	encephalalgia 46
diastole 73	duodenitis 111	encephalic 43, 46, 142
diastolic murmur 73	duodenum .. 98, 106, 109, 110, 112,	encephalitides 143
didymitis 132	113, 115, 119	encephalitis 43, 143
diencephalon 142	dysbolism 160	encephalodynia 46
differentiated 158	dysentery 19, 110	encephalogram 143
differentiation 52, 158	dyslexia 19	endemic 22
differentiation disorder 160	dyspepsia 19, 106	endocarditis 74
diffuse peritonitis 104	dyspeptic 106	endocardium 74
digest 98	dyspigmentation 160	endocrine 21
digestion 98	dysplasia 37, 158, 160	endocrine pancreas 145
digestive 98	dyspnea 90	endocrine part of pancreas 119
digital 15	dystocia 35, 137	endocrinologist 166
digital pad 155	dystrophia 33	endocrinology 166
digital phalanges 12, 56, 58	dystrophy 19, 33	endocytosis 21
dilatation of esophagus 103	dysuria 35	endometrial 43
dilation 66	**E**	endometriosis 35
dilator muscle 66		endometritis 43, 129, 137
dimer 26	ear 15, 43, 46, 151	endometrium 43, 128
diphyllobothriasis 163	earache 151	endomysium 64
diphyllobothrium 163	eccrine gland 154	endothelial cell 76
diplegia 36	electrocardiogram (ECG) 168	endotoxin 21
dipsesis 99	echocardiogram 70	enema 114

enteralgia ... 45, 110	esophagalgia ... 45	fascias ... 65
enterdynia ... 110	esophageal ... 45, 103	*Fasciola hepatica* ... 162
enteric ... 42, 45	esophageal adenocarcinoma ... 103	fascioliasis ... 162
enteritis ... 42, 110	esophageal diverticulum ... 103	fat necrosis ... 161
enteroanastomosis ... 38	esophageal obstruction ... 103	fatty change ... 160
enterocele ... 36	esophageal stenosis ... 103	fatty degeneration ... 160
enterodynia ... 45, 110	esophagectasia ... 103	fatty liver ... 115
enteroenterostomy ... 38	esophagi ... 103	feline ... 16
enteron ... 42, 45	esophagitis ... 103	feline acquired immunodeficiency syndrome ... 86
enteropexy ... 39	esophagodynia ... 45	
enterorrhaphy ... 38	esophagogastroanastomosis ... 38	feline immunodeficiency virus ... 86
entropion ... 150	esophagogastrostomy ... 38	feline infectious peritonitis ... 104
eosinocyte ... 80	esophagus ... 13, 45, 90, 94, 98, 103, 107	feline leukemia ... 81
eosinophil ... 80	ethnologist ... 166	feline leukemia virus ... 159
eosinophile ... 80	ethnology ... 166	feline panleukopenia ... 81
eosinophilia ... 34	ethologist ... 166	female dog ... 16
eosinophilic leukocyte ... 80	ethology ... 166	female goat ... 16
epicardia ... 106	eucaryote ... 50	female mouse ... 16
epicardium ... 74	euchromatin ... 19	femoral ... 15
epidemic ... 24	eugenic ... 19	femoral region ... 10
epidemiologist ... 166	eukaryote ... 50	femur ... 12, 56, 58
epidemiology ... 166	eutocia ... 137	fertility ... 136
epidermis ... 24, 154, 155	eutrophy ... 19	fertilization ... 128, 136
epididymal ... 133	ewe ... 16	fetlock joint ... 62
epididymides ... 133	excretory urography ... 125	fetus ... 136, 137
epididymis ... 132, 133	exocrine ... 21	fever reducer ... 168
epididymitis ... 134	exocrine part of pancreas ... 119	fibrinoid degeneration ... 160
epiglottis ... 90, 91	exocytosis ... 21	fibrinous inflammation ... 161
epilepsy ... 143	extension ... 66	fibrocartilage ... 60
epimenorrhea ... 35	extensor muscle ... 66	fibroma ... 159
epimysium ... 64	external ... 8	fibrosarcoma ... 159
epiphysial cartilage ... 59	external acoustic meatus ... 151	fibrous pericardium ... 74
epiphysis ... 59	external ear ... 43, 151	fibula ... 12, 56, 58
epistaxis ... 79, 91	external nose ... 91	filaria ... 163
epithelia ... 52	external urerthral orifice ... 129	filariasis ... 163
epithelial cell polarity ... 52	extracellular ... 21, 50	filiform ... 32
epithelial tissue ... 52	extrasystole ... 73	fill a cavity ... 100
epithelial tumor ... 159	extrauterine pregnancy ... 21, 129, 137	filly ... 16
epithelioma ... 159	exudative inflammation ... 161	final host ... 164
epithelium ... 24, 52	eye ... 15, 43, 46, 148	finger ... 15
epulis ... 101	eyelashes ... 149	flehmen ... 152
equine ... 16	eyelids ... 149	flesh ... 64
equine papilloma virus ... 159	**F**	flexion ... 66
erimysitis ... 64		flexor muscle ... 66
eroison and ulcer of abomasum ... 108	face ... 12	flies ... 164
erythrocyte ... 80	facial nerve ... 140	fluke ... 162
erythrocytopenia ... 36	fang ... 102	fly ... 164
erythropenia ... 36	fangs ... 102	foal ... 16
erythropoiesis ... 32	fascia ... 65	foot ... 12, 15
Escherichia coli ... 113	fasciae ... 65	foot-and-mouth-disease ... 99

177

forearm ·················· 12, 15	gastrointestinal tract ·········· 110	granulomatous inflammation ·· 161
forearm region ················ 10	gastroptosis ·················· 36	gray matter ················ 141
foreign body ················ 161	gastroscope ·················· 30	gray substance ·············· 141
forelimb ····················· 15	gastroscopy ·················· 30	greater curvature ············ 105
foremilk ···················· 131	gene ························ 53	gums ······················ 100
forensic medicine ············ 166	gene therapy ················· 53	gustation ··················· 153
fourth ventricle ·············· 142	genetically modified organism ·· 53	gustatory ··················· 153
frequent urination ············ 125	geneticist ············ 33, 53, 166	gustatory bud ··············· 153
frog ···················· 155, 156	genetics ············· 33, 53, 166	guttural pouch ··············· 151
frontal ······················· 9	genicular ····················· 15	gynecologist ················· 166
frontal sinus ·················· 90	genome ······················ 53	gynecology ·················· 166
fundal hemorrhage ··········· 150	germicide ···················· 34	
fundus of stomach ······ 105, 106	gerontologist ················ 166	# H
fungiform ···················· 32	gerontology ················· 166	
funiculus spermaticus ········ 134	giardia ······················ 162	hair ························ 154
	giardiasis ··················· 162	hair follicle ·················· 154
# G	gingiva ················· 100, 101	hair loss ···················· 168
	gingivae ···················· 101	hair papilla ·················· 154
galactopoietic ··············· 131	gingival ····················· 101	hammer ····················· 151
galactorrhea ·················· 35	gingivitis ···················· 101	hand ························ 15
galactosamine ··············· 131	GI tract ····················· 110	head ························ 15
galactose ···················· 131	gland ···················· 46, 52	head of muscle ··············· 65
gall ························ 116	glandular parts ··············· 106	heart ········ 13, 14, 42, 45, 70, 94
gall bladder ·········· 14, 42, 117	glaucoma ················ 25, 150	heart attack ·················· 72
gall bladder cancer ··········· 118	glial cell ···················· 141	heart block ··················· 73
gallbladder ·············· 115, 117	gliocyte ····················· 141	heart failure ·················· 70
gallbladder carcinoma ········ 118	glomerular capsule ··········· 122	heart graft ···················· 72
gallbladder sludge ··········· 118	glomerulo-nephritis ··········· 123	heart murmur ················· 73
gallstone ···················· 118	glossal ················· 42, 45, 153	heel ························· 61
ganglion cell ················· 141	glossalgia ··············· 45, 153	hemagglutination inhibition test ·· 78
gangrene ···················· 161	glossitis ············· 42, 99, 153	hemal ······················ 78
gangrenous inflammation ····· 161	glossodynia ············· 45, 153	hemal lymph node ············ 78
gap junction ·················· 50	glossopharyngeal nerve ······ 140	hemangiopericytoma ·········· 75
gastralgia ················ 45, 105	glucagon ···················· 119	hemangiosarcoma ············ 75
gastrectomy ················· 105	gluteal ······················ 15	hematic ····················· 78
gastric ············ 14, 42, 45, 105	gluteal region ············ 10, 15	hematocele ··················· 36
gastric cancer ··············· 105	glycogen granules ············ 50	hematocrit ··················· 78
gastric dilatation ············· 105	glycopenia ··················· 36	hematogenesis ··············· 81
gastric dilation volvulus ······ 105	goat ························ 16	hematologist ················· 166
gastric flu ··················· 105	goats ······················· 16	hematology ·················· 166
gastric fluid ·················· 98	goiter ······················ 145	hematopoiesis ············ 32, 81
gastric irrigation ·············· 105	gonad ······················ 145	hematopoietic tissue ·········· 81
gastric juice ·················· 98	Gorgi apparatus ··············· 51	hematopyuria ················· 44
gastric lavage ················ 105	gout ······················· 160	hematothorax ················ 78
gastric torsion ················ 105	graft ························ 86	hematouria ··················· 78
gastric ulcer ················· 105	granular layer ················ 155	hematuria ··················· 125
gastritis ················· 42, 105	granular leukocyte ············ 80	hemic ······················ 78
gastroanastomosis ············ 38	granulation tissues ············ 161	hemiplegia ··················· 36
gastrocamera ················ 105	granule cell ················· 155	hemisection ·················· 20
gastrodynia ············· 45, 105	granulocyte ·················· 80	hemoglobin ············· 78, 160
gastroenteritis ················ 105		hemolymph node ············· 78

hemolysis · · · · · · · · · · · · · · · 35, 78	histology · · · · · · · · · · · · · · · · · · · 166	hyperthyroidism · · · · · · · · · · 33, 145
hemolytic anemia · · · · · · · · · · · 78	hock · 12	hypertrophia · · · · · · · · · · · · · · · · · · 33
hemophilia · · · · · · · · · · · · · · · 34, 78	hock joint · · · · · · · · · · · · · · · · · · · 62	hypertrophy · · · · · · · · · · · · · · 33, 161
hemopoiesis · · · · · · · · · · · · · · · · · 81	hog · 16	hypocalcemia · · · · · · · · · · · · · · · 160
hemorrhage · · · · · · · · · · · · · · · · · 79	homeostasis · · · · · · · · · · · · · · · · · 24	hypoglycaemia · · · · · · · · · · · · · · · 79
hemorrhage in the eyegrounds 150	homogeneous · · · · · · · · · · · · · · · · 24	hypoglycemia · · · · · · · · · · · · · · · · 79
hemorrhagic diseases · · · · · · · 78	hoof · 156	hypoglossal nerve · · · · · · · · · · 140
hemorrhagic inflammation · · · · 161	hooves · 156	hypoinsulinism · · · · · · · · · · · · · · · 33
hemorrhoid · · · · · · · · · · · · · · · · · 114	horn · 15, 155	hypomania · · · · · · · · · · · · · · · · · · 34
hemosiderin · · · · · · · · · · · · · · · · 160	horse · 16	hypoparathyroidism · · · · · · · · · · 145
hemothorax · · · · · · · · · · · · · · 78, 92	horses · 16	hypophyseal · · · · · · · · · · · · · · · · 144
Henle's loop · · · · · · · · · · · · · · · 122	human · 16	hypophysial · · · · · · · · · · · · · · · · · 144
heparin · 115	human immunodeficiency virus · · 86	hypophysis · · · · · · · · · · · · · · · · · 144
hepatalgia · · · · · · · · · · · · · · · · · · 45	humans · 16	hypoplasia · · · · · · · · · · · · · · · · · · 37
hepatic · · · · · · · · · 14, 42, 45, 115	humerus · · · · · · · · · · · · · · 12, 56, 58	hypotension · · · · · · · · · · · · · · 23, 79
hepatic coma · · · · · · · · · · · · · · · 115	humoral immunity · · · · · · · · · · · · 85	hypothalamus · · · · · · · · · · · · 23, 142
hepatic duct · · · · · · · · · · · · · · · · 115	hyaline cartilage · · · · · · · · · · · · · 60	hypothyroidism · · · · · · · · · · · · · · 145
hepatic failure · · · · · · · · · · · · · · 115	hyaline degeneration · · · · · · · · 160	hypouresis · · · · · · · · · · · · · · · · · 125
hepatic lobule · · · · · · · · · · · · · · 115	hyaline droplet degeneration · 160	hypoxia · 78
hepatic steatosis · · · · · · · · · · · 115	hydrocephalus · · · · · · · · · · · · · · 143	hysteralgia · · · · · · · · · · · · · · · 46, 129
hepatic triad · · · · · · · · · · · · · · · · · 26	hydrochloric acid · · · · · · · · · · · 106	hysterectomy · · · · · · · · · · · · · · · 129
hepatitis · 42	hydrometra · · · · · · · · · · · · · · · · · 129	hysterodynia · · · · · · · · · · · · 46, 129
hepatocyte · · · · · · · · · · · · · · · · · 115	hydronephrosis · · · · · · · · · · · · · 123	hysterogram · · · · · · · · · · · · · · · · 129
hepatoid gland carcinoma · · · · 114	hydropericardium · · · · · · · · · · · · 74	hysterography · · · · · · · · · · · · · · 129
hepatolith · · · · · · · · · · · · · · · · · · 118	hydroperitoneum · · · · · · · · · · · 104	hysterorrhexis · · · · · · · · · · · · · · 137
hepatolithiasis · · · · · · · · · · · · · · 118	hydrophobia · · · · · · · · · · · · · · · · · 34	hysteroscope · · · · · · · · · · · · · · · 129
hepatomegaly · · · · · · · · · · · · · · 115	hydropic degeneration · · · · · · · 160	hysterotomy · · · · · · · · · · · · · · · · · 37
hepatophyma · · · · · · · · · · · · · · · 115	hydrosalpinx · · · · · · · · · · · · · · · · 128	
hernia · 35	hydrothorax · · · · · · · · · · · · · · · · · 92	I
hernia of intervertebral disk · · · · 63	hygiene · 166	ichthyologist · · · · · · · · · · · · · · · 166
herpes virus · · · · · · · · · · · · · · · 159	hygienist · · · · · · · · · · · · · · · · · · · 166	ichthyology · · · · · · · · · · · · · · · · · 166
heterochromatin · · · · · · · · · · · · · 24	hymen · 129	icterus · 115
heterogeneous · · · · · · · · · · · · · · 24	hymenectomy · · · · · · · · · · · · · · · · 37	ileac · 111
hiccup · 93	hymenotomy · · · · · · · · · · · · · · · · 37	ileal · 111
high blood glucose level · · · · · · 79	hyoid bone · · · · · · · · · · · · · · · · · · 90	ileitis · 111
high blood pressure · · · · · · · · · · 79	hyperacid · · · · · · · · · · · · · · · · · · 106	ileum · · · · 109, 111, 112, 113, 119
himantosis · · · · · · · · · · · · · · · · · · 91	hyperacidity · · · · · · · · · · · · · · 23, 106	ileums · 111
hindlimb · 15	hyperadrenocorticism · · · · · · · 145	ilio-femoral lymph node · · · · · · · 82
hindpaw · 58	hypercalcemia · · · · · · · · · · 78, 160	ilium · 58
hip · 15	hyperemia · · · · · · · · · · · · · · · · · · 78	immature · · · · · · · · · · · · · · · · · · · 52
hip bone · · · · · · · · · · · · · · · · 56, 58	hyperestrinism · · · · · · · · · · · · · 145	immortal · 19
hip dysplasia · · · · · · · · · · · · 59, 63	hyperfunction of pituitary · · · · · 144	immune · 85
hip joint · · · · · · · · · · · · · 46, 61, 62	hypergalactia · · · · · · · · · · · · · · · 131	immune response · · · · · · · · · · · 85
hip joint region · · · · · · · · · · · · · · · 10	hyperglycaemia, hyperglycemia · · 79	immunization · · · · · · · · · · · · · · · · 85
hip region · · · · · · · · · · · · · · · · · · · 10	hyperparathyoridism · · · · · · · · 145	immunodeficiency · · · · · · · · · · · 85
His bundle · · · · · · · · · · · · · · · · · · 73	hyperparathyreodismus · · · · · · 145	immunoenhancement · · · · · · · · 85
histiocyte · · · · · · · · · · · · · · · · · · · 81	hyperpigmentation · · · · · · · · · · 160	immunofluorescence · · · · · · · · · 85
histiocytoma · · · · · · · · · · · · 81, 159	hyperpituitarism · · · · · · · · · · · · · 144	immunogenicity · · · · · · · · · · · · · 85
histocyte · · · · · · · · · · · · · · · · · · · 81	hyperplasia · · · · · · · · · · · · · · 37, 161	immunoglobulin · · · · · · · · · · · · · 85
histologist · · · · · · · · · · · · · · · · · 166	hypertension · · · · · · · · · · · · · · 23, 79	immunohistochemistry · · · · · · · 85

immunologist · · · · · · · · · · · · · · · · 166	intervertebral disk protrusion · · 63	**K**
immunology · · · · · · · · · · · · · · · 166	intervertebral foramen · · · · · · · · 57	
immunoprecipitation · · · · · · · · · · 85	intestinal · · · · · · · 42, 45, 109, 110	kaliopenia · 36
immunosuppression · · · · · · · · · · 85	intestinal anastomosis · · · · · · · · 110	Kaposi sarcoma · · · · · · · · · · · · · · 86
implantation · · · · · · · · · · · · 128, 137	intestinal flora · · · · · · · · · · · · · · 110	karyolysis · 50
in vitro fertilization · · · · · · · · · · · · 136	intestinal fluid · · · · · · · · · · · · 98, 110	karyoplast · 50
incisor · · · · · · · · · · · · · · · · · 90, 102	intestinal juice · · · · · · · · · · · · · · · · 98	karyotheca · · · · · · · · · · · · · · · · · · 50
inclusion body formation · · · · · · 161	intestine · · · · · · · · · · · · 42, 45, 109	keratic · 154
incus · 151	intracellular · · · · · · · · · · · · · · 21, 50	keratin degeneration · · · · · · · · · 160
indigested · 98	intraepidermal macrophage · · 155	keratinized cell · · · · · · · · · · · · · · · 155
indigestibility · · · · · · · · · · · · · · · · · 98	intramuscular（I.M.）· · · · · · · · · 168	keratitis · · · · · · · · · · · · · · · · 43, 150
indigestible · · · · · · · · · · · · · · · · · · · 98	intravascular hemolysis · · · · · · · · 75	keratoma · · · · · · · · · · · · · · · · · · · 155
indigestion · · · · · · · · · · · · · · · · · · · 98	intravenous drip · · · · · · · · · · · · · 168	keratorrhexis · · · · · · · · · · · · · · · · · · 36
infectious canine hepatitis · · · · 115	intravenous injection · · · · · · · · · · 21	keratosis · 155
infectious coryza · · · · · · · · 91, 152	intravenous（I.V.）· · · · · · · · · · · 168	ketosis · 35
inferior · 9	intussusception · · · · · · · · · · · · · · 110	kid · 16
Inflammation · · · · · · · · · · · · · · · · 161	involuntary muscle · · · · · · · · · · · · 65	kidney · · 13, 14, 42, 46, 122, 127, 132, 144
inflammatory bowel disease · · 110	inward rotation · · · · · · · · · · · · · · · · 66	kidney cancer · · · · · · · · · · · · · · · 123
infraorbital nerve · · · · · · · · · · · · · · 24	iridesis · 150	kidney stones · · · · · · · · · · · · · · · 125
infraspinatus muscle · · · · · · · · · · 24	iridal · 149	kidney transplant · · · · · · · · · · · · 123
infusion · 86	iridial · 149	kitten · 16
ingest · 98	iridian · 149	knee · 12, 15
ingestion · 98	iridic · 149	knee joint · · · · · · · · · · · · · · · · 61, 62
inguinal canal · · · · · · · · · · · · · · · 133	iris · 148, 149	**L**
inguinal lymph node · · · · · · · · · · 82	iron deficiency · · · · · · · · · · · · · · · · 78	
inject · 168	irregular heartbeat · · · · · · · · · · · · 73	labor pains · · · · · · · · · · · · · · · · · 137
injection · 168	ischemia · 78	laboratory examination · · · · · · 168
inner ear · 151	ischialgia · 46	laceration · · · · · · · · · · · · · · · · · · · 66
insane · 19	ischioneuralgia · · · · · · · · · · · · · · · 46	lacrimal gland · · · · · · · · · · · · · · · 149
insecticide · 34	ischium · 58	lactation · 131
insertion · 65	islets of Langerhans · · · · · · · · · · 119	lactic · 131
insomnia · 19	I.V. drip · 168	lactic acid · · · · · · · · · · · · · · · · · · · 131
inspection · · · · · · · · · · · · · · · · · · · 168	ixodidae · 164	lactiferous · · · · · · · · · · · · · · · · · · · 131
insulin · 119	ixodosis · 164	lactiferous duct · · · · · · · · · · · · · · 131
insulin secreting islet cell tumor · · 145	**J**	lactiferous sinus · · · · · · · · · · · · · 131
insulinoma · · · · · · · · · · · · · · · · · · 145		lactose · 131
intercalated disc · · · · · · · · · · · · · · · 21	Japanese encephalitis · · · · · · · · 143	lamb · 16
intermediate · · · · · · · · · · · · · · · · · · 8	jaundice · · · · · · · · · · · · · · · 115, 160	Langerhans cell · · · · · · · · · · · · · 155
intermediate host · · · · · · · · · · · · 164	jaw · 61	laparocele · · · · · · · · · · · · · · · · · · 104
internal · 8	jejunal · 111	laparotomy · · · · · · · · · · · · · · · · · 104
internal bleeding · · · · · · · · · · · · · · 79	jejunitis · 111	large bowel · · · · · · · · · · · · · · · · 110
internal ear · · · · · · · · · · · · · · 43, 151	jejunum · · · · · · 111, 112, 113, 119	large intestine · · 98, 110, 112, 113
internal elastic membrane · · · · · · 76	jejunum and ileum · · · · · · · · 13, 98	laryngalgia · · · · · · · · · · · · · · · · · · 45
internal medicine · · · · · · · · · · · · 166	jejunums · 111	laryngeal · · · · · · · · · · · · · · 42, 45, 91
internist · 166	joint · · · · · · · · · · · · · · · · · · 43, 46, 62	laryngeal paralysis · · · · · · · · · · · · 91
interstitial cell tumor · · · · · 132, 145	joint capsule · · · · · · · · · · · · · · · · · 63	laryngitis · · · · · · · · · · · · · · · · · 42, 91
interstitial fatty infiltration · · · · · · 160	joint cavity · · · · · · · · · · · · · · · · · · · 63	laryngohemiplegia · · · · · · · · · · · · 91
interstitial peumonia · · · · · · · · · · 95	joint mice · 63	larynx · · · · · · · · · · · · · · · · 42, 45, 91
intervertebral disc herniation · · 143	joint mouse · · · · · · · · · · · · · · · · · · 63	lateral · 8, 11
intervertebral disk · · · · · · · · · · · · · 63	jugular vein · · · · · · · · · · · · · · · · · · 77	lateral laminae · · · · · · · · · · · · · · 131

lateral ventricle ············ 142	lumbar vertebrae ············ 57	malignant lymphoma ········ 159
laxative ················· 168	luminal ·················· 52	malignant melanoma ········ 159
left atrio-ventricular valve ······ 71	lung ············ 14, 42, 94, 95	malignant seminoma ········ 159
left atrium ················ 71	lung abscess ·············· 95	malleus ·················· 151
left bundle branches ········· 73	lung cancer ··············· 95	mamillary process ··········· 57
left limb ·················· 73	lungworm ················· 95	mamma ················ 127, 130
left lung ················ 93, 94	luteal hypoplasia ·········· 127	mammal ··············· 46, 130
left ventricle ············ 14, 71	luxation (LX) ············· 170	mammalgia ············· 46, 130
left ventricular ············· 14	lympadenectomy ············ 82	mammaplasty ·············· 37
leg ······················ 12	lympadenosis ·············· 82	mammary ··············· 43, 46
legal medicine ············ 166	lymph ···················· 82	mammary adenocarcinoma ·· 130
leiomyoma ················ 159	lymph capillary ·········· 75, 82	mammary gland ···· 43, 130, 131
leiomyosarcoma ············ 159	lymph corpuscle ············ 80	mammary glands ··········· 127
lens ···················· 148	lymph node ················ 82	mammary tumor ············ 130
lentiform ·················· 32	lymph nodule ·············· 82	mammectomy ·············· 130
lesser curvature ··········· 105	lymph vessel ··············· 82	mammography ············· 130
leucopenia ················ 81	lymphaden ················· 82	mammoplasty ·············· 37
leukemia ··············· 78, 159	lymphadenitis ·············· 82	man ······················ 16
leukocyte ············· 80, 81	lymphadenopathy ············ 82	mandible ·················· 12
leukocyte adhesion deficiency ·· 81	lymphangiectasis ············ 82	mandibular lymph node ······· 82
leukocytopenia ············· 36	lymphangioma ·············· 82	manual ··················· 15
leukopenia ················ 36	lymphangitis ··············· 82	mare ····················· 16
levator muscle ············· 66	lymphatic ················· 82	Marek's disease ············ 159
Leydig cell tumor ······· 132, 145	lymphatic corpuscle ·········· 80	margo plicatus ············· 106
life cycle ················ 164	lymphatic vessel ········ 75, 82	massage ················· 169
lifting up ················· 66	lymphocyte ·········· 80, 81, 82	mast cell ·············· 81, 85
ligament ·················· 65	lymphoduct ················ 82	mast cell tumor ············· 81
linea alba ·················· 25	lymphoma ··············· 81, 82	mastadenoma ·············· 130
lingual ········ 14, 42, 45, 153	lymphonode ················ 82	mastalgia ··············· 46, 130
lingual epithelium ·········· 153	lymphosarcoma ············· 82	mastectomy ············ 37, 130
linguopapilitis ············· 153	lysosome ·················· 51	masticate ················· 100
lion ······················ 12		mastication ··············· 100
lipid and lipoid dysbolism ···· 160	# M	mastitis ··············· 43, 130
lipid droplet ··············· 50	machinery murmur ··········· 73	mastocyte ················· 85
lipidosis ················· 160	macromolecule ············· 25	mastodynia ············· 46, 130
lipofuscin ················ 160	macrophage ············ 25, 85	mastotomy ················· 37
lipoid ····················· 32	macula adherens ············ 50	mating ··················· 136
lipoma ··················· 159	main olfactory bulb ········· 152	matricial ·················· 43
liposarcoma ··············· 159	major duodenal papilla ······ 115	matrix unguis ·············· 155
lipotrophy ················· 33	major histocompatibility complex ·· 86	maturation ················· 52
liquefactive necrosis ········ 161	malacia ·················· 161	mature ···················· 52
liver ·· 13, 14, 42, 45, 98, 109, 115, 117	malaria ··················· 35	measles ··················· 95
liver cancer ··············· 115	male cat ·················· 16	mechanical ileus ············ 111
liver fluke ············ 115, 162	male dog ·················· 16	medial ·················· 8, 11
loin ······················ 15	male goat ·················· 16	medial iliac lymph node ······· 82
loss of one's eyesight ········ 150	male mouse ················ 16	medial laminae ············ 131
low blood glucose level ······· 79	malformation ············ 19, 161	median ···················· 8
low blood pressure ··········· 79	malignant ··············· 19, 159	mediastinal emphysema ······ 92
lower leg ·················· 15	malignant fibrous histiocytoma ·· 81	mediastinal lymph node ······· 82
lumbar ···················· 15	malignant histiocytoma ······ 159	mediastinum ··············· 92

181

medical attorney 166	midsagittal 8	muscular 43, 46, 64
medulla 122	milk 131	muscular dystrophy 66
medulla oblongata 142	milk fever 131	muscular rheumatism 66
megacolon 25, 113	milk line 131	muscular tissue 52
megaesophagus 103	milk teeth 102	muzzle 10, 91
megakaryocyte 25, 81	milking 131	myalgia 46
meibomian tumor 159	milky 131	myasthenia gravis 66
meiosis 52	mineral dysbolism 160	mycologist 166
melanin 25, 160	miscarriage 19, 137	mycology 166
melanocyte 155	misdiagnosis 19	myelalgia 143
melanoma 25	mitochondria 51	myelatrophy 143
men 16	mitosis 52	myelic 59, 143
meningioma 159	mitral incompetence 72	myelitis 59, 143
menorrhalgia 46	mitral insufficiency 72	myeloblast 59
menstrual 46	mitral regurgitation 72	myeloblastoma 59
menstrualtion 46	mitral valve 71	myeloid 32, 143
mesangium 23, 75	molar 102	myeloid tissue 81
mesencephalon 23, 142	molecular biologist 166	myeloma 59
mesenchyme 23	molecular biology 166	myelopoiesis 32
mesentery 110	monoclonal 26	myelosis 59
mesoderm 23	monocyte 80, 81	myiasis 164
messenger RNA (mRNA) 53	monocytosis 81	myoatrophy 64
metacarpal bones 12, 56, 58	monomania 34	myoblast 64
metacarpal region 10	monomer 26	myocardia 70
metacarpophalangeal joints 61	morphologist 166	myocardial infarction 72
metanephros 23	morphology 166	myocarditis 64, 70
metaphysis 59	morphometrics 33	myocardium 70
metaplasia 160	motoneuron 140	myocyte 64
metastases 158	motor nerve 140	myodynia 46
metastasis 158	mouse 16	myofibril 64, 65
metastasize 158	mouth 15, 45, 98, 99	myoglobin 160
metatarsal bones 12, 56, 58	magnetic resonance imaging (MRI) 168	myoglobulin 64
metatarsus 23	mucinous degeneration 160	myology 64
metencephalon 23	mucometra 129	myoma 64
metrical 43	mucosa 52	myometritis 129
metritis 43, 129	multiple fission 164	myopathy 35, 64, 66
metroptosis 36	multiple myeloma 59, 159	myorrhexis 36
mice 16	multiple sclerosis 86	myositis 43, 64, 66
microanastomosis 38	multiplication 52	myxoma 159
microbiologist 33, 166	multivesicular body 26	
microbiology 25, 166	murine 16	# N
microfilament 51	muscarinic receptor 141	nail 156
microglia 141	muscle 43, 46, 64	nail bed 155
microscope 30	muscle and connective tissue .. 151	narcolepsy 143
microscopy 25, 30	muscle belly 65	nares 91
microtubule 51	muscle contusion 66	naris 91
microvilli 51	muscle contracture 66	nasal 15, 42, 45, 152
midbrain 142	muscle fiber 65	nasal bleeding 79
middle ear 43, 151	muscle strain 170	nasal cavity 90, 91
middle phalanx 58, 62	muscle tail 65	nasal discharge 91

nasal planum 155	nexus 50	olfactory bulb 142
navel 15	nictitating membrane 149	olfactory epithelium 152
nebula 150	night blindness 150	olfactory nerves 140
neck 12, 15	nipple 131	oligodendrocyte 141
neck of tooth 100	no heartbeat 73	oligodendroglia 141
necrosis 52, 161	non-alcoholic steatohepatitis .. 115	oligodendroglioma 159
necrozoospermia 135	non-epithelial tumor 159	oligozoospermia 135
nematode 163	non-glandular part 106	oliguria 125
neogala 131	nonimmune 19	omasal impaction 108
nephralgia 46, 123	noninvasive 19	omasitis 108
nephrectomy 123	nonpathogenic 19	omasum 107, 108
nephritis 42, 123	nonunion 63	oncogene 158
nephrolith 125	nose 15, 42, 45, 152	oncogenesis 158
nephron 122	nose bleeding 91	oncologist 167
nephron loop 122	nostril 91	oncology 158, 167
nephropathy 35	nourishment 98	oncovirus 158
nephropexy 39	nuclear 50	oogenesis 32
nephrosclerosis 123	nuclear envelope 50	oophoralgia 127
nephrosis 123	nuclear matrix 50	oophorectomy 127
nerve 46, 140	nuclear membrane 50	oophoritis 127
nerve cell 140, 141	nucleic acid dysbolism 160	oophoroma 127
nerve fiber 152, 154	nucleolus 50	opening of vomeronasal organ .. 152
nerve fibers 153	nucleoplasm 50	ophthalmalgia 46, 148
nervi 140	nucleus 50	ophthalmia 148
nervous 46, 140	nutmeg liver 115	ophthalmic 46, 148
nervous tissue 52	nyctalopia 150	ophthalmitis 43
neuralgia 46, 140	nymphomania 34	ophthalmologist 167
neuritides 140		ophthalmology 167
neuritis 140	**O**	ophthalmopathy 148
neurodynia 46, 140	obesity 168	opportunistic infection 86
neurogenesis 32	obstetrician 166	optic 15, 43, 148
neurogenic shock 140	obstetrics 166	optic nerve 140, 148
neurohypophysis 144	obstruction 76	opticokinetic nystagmus 148
neurologist 166	obstruction of omasum 108	optokinetic nystagmus 148
neurology 140, 166	OCD lesion 63, 170	optometrist 148
neuroma 140	ocular 15, 46, 148	oral 15, 45, 99
neuromodulator 141	oculodynia 46, 148	oral point 99
neuromuscular electrical stimulation	oculomotor nerve 140	orale 99
(NMES) 169	oculopathy 148	orchialgia 46, 132
neuron 140, 141	odontalgia 45, 100	orchiodynia 46, 132
neuropathy 35, 140	odontitis 42, 100	orchioncus 132
neuroplegia 36	odontodynia 45, 100	orchiopexy 132
neurosis 140	odontologist 167	orchis 132
neurotransmitter 141	odontology 100, 167	orchitis 132
neurotrophy 33	odontoma 100	organelles 50
neuter 137	olecranon region 10	organization 161
neutropenia 36	olfaction 152	origin 65
neutrophil 80, 85	olfactometry 152	ornithologist 167
neutrophile 80	olfactophobia 152	ornithology 167
neutrophilic 80	olfactory 152	orthodontics 19

183

orthopedics	167	
orthopedics	19	
orthopedist	167	
oscheal	134	
oscillograph	31	
osseous	46	
osseous tissue	52	
ossify	56	
ostalgia	46	
osteal	43, 46, 56	
ostealgia	56	
osteitis	43, 56	
osteoarthritis	57, 63	
osteoblast	56	
osteochondritis dissecans	63, 170	
osteochondrosis dissecans	56	
osteoclast	56	
osteocyte	56	
osteodynia	46, 56	
osteodystrophy	56	
osteolysis	35, 56	
osteoma	56, 159	
osteomalacia	56	
osteomyelitis	56, 59	
osteon	56	
osteoporosis	56	
osteopsathyrosis	57	
osteotomy	56	
otalgia	46	
otic	15, 43, 46, 151	
otitis	43, 151	
otitis externa	43, 151	
otitis interna	43, 151	
otitis media	43, 151	
otoacariasis	151	
otodynia	46	
otorhinolaryngologist	167	
otorhinolaryngology	167	
outward rotation	66	
ovarialgia	127	
ovarian	14, 127	
ovarian adhesion	127	
ovarian cyst	127	
ovariectomy	127	
ovariohysterectomy	37, 127, 137	
ovarioncus	127	
ovariotomy	37	
ovaritis	127	
ovary	13, 14, 127, 144	
oviduct	42, 127, 128	
ovine	16	
ovulation	128, 136	
ovulation failure	127	
ovum pick up	136	
ox	16	
oxen	16	
oxyphil	80	
oxyphile	80	
oxyphilic leukocyte	80	
oxyuriasis	163	
oxyurid	163	

P

pad	155
painkiller	168
palatine tonsil	83
palatoplasty	37
palatorrhapy	38
paleontologist	167
paleontology	167
palmar	11
palpation	168
panangiitis	24
pancreas	14, 42, 46, 98, 109, 115, 119
pancreatalgia	46, 119
pancreatectomy	119
pancreatic	14, 42, 46, 119
pancreatic calculus	119
pancreatic cancer	119
pancreatic duct	115, 119
pancreatic fluid	98
pancreatic islets	119, 144
pancreatic juice	98
pancreatic polypeptide	119
pancreatic vein	119
pancreaticoduodenal	119
pancreaticosplenic lymph nodes	119
pancreatitis	42, 119
pancreatolith	118, 119
pancreatolithiasis	119
pancreolith	119
pancreozymin	119
pandemic	24
panneuritis	24
panostitis	56
papillary duct	131
papillary muscles	71
papilloma	159
papova virus	159
paracentesis	38
paracrine	21
paragonimiasis westermani	162
paragonimus westermani	162
paralysis	35, 143
paralysis of esophagus	103
paralysis of urinary bladder	124
paralytic ileus	111
paralyze	21
paranasal sinuses	91
paraplegia	36
parasite	21, 162
parasitologist	167
parasitology	167
parastruma	145
parasympathetic nerve	140
parathyroid gland	144
parenchyma	158
parenchymatous inflammation	161
parietal layer of the serous pericardium	74
parotid lymph node	82
parotitis	99
parvocellular	25
parvovirus	25
pastern bone	58, 62
pastern joint	62
patella	56, 58
patellar luxation	59
patellar region	10
patent ductus arteriosus	72
patent foramen ovale	72
pathological examination	168
pathologist	167
pathology	167
paw	10
pectoral	15
pectoral region	10
pedal	15
pedal phalanges	12, 56, 58
pediatrician	167
pediatrics	167
pedophilia	34
pelvic	104
pelvic limb	15
pelvis	104
penes	134
penile	46, 134
penis	13, 46, 132, 133, 134
pepsin	106
peptic	106
peptic ulcer	106

peptic ulcer disease 106	pharyngeal 103	pneumonopexy 39
peptide 53	pharyngeal paralysis 103	pneumothorax 92, 95
per os (P.O.) 168	pharyngitis 103	pneumovagina 130
percussion 168	pharynx 98, 103	poll or nape of the neck 12
percutaneous 22	pheromone 152	pollakiuria 125
perianal fistula 114	phobism 33	polyclonal 26
perianal gland tumor 114	phonocardiogram 70	polydipsia 99
pericardial cavity 74	phrenic 93	polymenorrhea 35
pericardial effusion 74	phrenic nerve 93	polymer 26
pericardial fluid 74	phrenitis 93	polymyositis 66
pericarditis 74	physical examination 168	polyp 159
pericardium 22	physiologist 167	polyps of rectum 113
perichondritis 60	physiology 167	polyradiculoneuritis 143
perichondrium 60	physiotherapy 39	polyuria 125
perimetritis 129	pig 16	pons 142
perimyositis 64	piglet 16	poor appetite 168
perimysiitis 64	pigment granule 50	popliteal lymph node 82
perimysium 64	pigs 16	popliteal region 10
periodontal 42	pineal gland 142, 144	porcine 16
periodontal disease 100	pinealoma 145	portal vein 77, 79
periodontal tissue 42	pinna 151	portsystemic shunt 76
periodontitis 42, 100	piriform 32	posterior 9, 23
periodontium 100	piroplasma 162	posterior chamber of eye ... 148
periosteal reaction 57	piroplasmosis 162	postmortem 23
periosteum 22, 56	pituitary 144	postmortem change 161
peripheral 11	pituitary gigantism 144	postmortem dots 161
peripheral nervous system 140, 141	pituitary gland 142, 144	postnatal 23
peritoneal 42, 46, 104	placenta 128, 136, 137	PP cell 119
peritonealgia 46	plantar 11	precancerous 23
peritoneocentesis 38	plasma 80	premature birth 137
peritoneopericardial diaphragmatic hernia 104	plasma cell 85	premature complex 73
peritoneoscopy 104	plasmacyte 85	premolar 102
peritoneum 42, 46, 104	plastic surgeon 167	prenatal 23
peritonitis 42, 104	plastic surgery 167	prescription diet 168
permanent teeth 102	platelet 80	prickle cell 155
peroral 22	pleura 42, 45, 92	primary hair 154
persistent estrus 136	pleural 42, 45, 92	proctalgia 45
persistent right aortic arch 72	pleural effusion 92	proctectomy 114
pesticide 34	pleuralgia 45	proctitis 113
positron emission tomography (PET) .. 168	pleurisy 92	proctocele 114
Peyer's patch 82	pleuritis 42, 92	proctodynia 45
phagocytize 85	pleurocentesis 38	proctologist 167
phagocytose 85	pleurodynia 45	proctology 167
phagocytosis 85	plumonary lobectomy 95	proctoptosis 36
phallalgia 46	pneocardiac reflex 90	proctostomy 114
phallodynia 46	pneumatosis 90	profundus 11
pharmacist 167	pneumoconiosis 95	prognosis 23
pharmacologist 167	*Pneumocystis carinii* pneumonia .. 86	programmed cell death 52
pharmacology 167	pneumomediastinum 92	progressive change 161
	pneumonia 35, 42, 95	progressive neurodeatrophia .. 150

185

proliferation · · · · · · · · · · · · · · · · · 52	pulmonary insufficiency · · · · · · · 72	recta · 112
proliferative inflammation · · · · 161	pulmonary stenosis · · · · · · · · · · · 72	rectal · 112
prolonged estrus · · · · · · · · · · · 136	pulmonary tuberculosis · · · · · · · · 95	rectal cancer · · · · · · · · · · · · · · · 113
pronation · · · · · · · · · · · · · · · · · · · 66	pulmonary valve · · · · · · · · · 71, 95	rectal palpation · · · · · · · · · · · · · 136
pronator muscle · · · · · · · · · · · · · 66	pulmonary vein · · · · · · · · · · · · · · 71	rectal prolapse · · · · · · · · · · · · · · 113
proper gastric gland · · · · · · · · · 105	pulmonic incompetence · · · · · · · · 72	rectitis · 113
prostata · 133	pulpal · 42	rectocele · · · · · · · · · · · · · · · · · · 114
prostatalgia · · · · · · · · · · · · · · · · 134	pulpitis · · · · · · · · · · · · · · · · · · 42, 100	rectum · · 13, 45, 98, 109, 112, 113, 133
prostate · · · · · · · · · · · · 23, 132, 133	punctiform · · · · · · · · · · · · · · · · · · 32	rectums · · · · · · · · · · · · · · · · · · · 112
prostate gland · · · · · · · · · · · · · · 133	puppy · 16	red blood cell · · · · · · · · · · · · · · · 80
prostatic · · · · · · · · · · · · · · · · · · · 133	purulent inflammation · · · · · · · · 161	red corpuscle · · · · · · · · · · · · · · · 80
prostatic hypertrophy · · · · · · · · 134	putrefaction · · · · · · · · · · · · · · · · 161	red pulp · · · · · · · · · · · · · · · · · · · 84
prostatitis · · · · · · · · · · · · · · · · · · 134	pyelitis · 123	reflux esophagitis · · · · · · · · · · · 103
prostatodynia · · · · · · · · · · · · · · · 134	pyelonephritis · · · · · · · · · · · · · · · 123	regeneration · · · · · · · · · · · · 52, 161
prostatomegaly · · · · · · · · · · · · · 134	pyemia · 44	region of elbow joint · · · · · · · · · · 10
protein dysbolism · · · · · · · · · · · 160	pyloric canal · · · · · · · · · · · · · · · · 106	region of thigh · · · · · · · · · · · · · · · 10
proteolysis · · · · · · · · · · · · · · · · · · 35	pyloric dysfunction · · · · · · · · · · 106	rejection · 86
proteomics · · · · · · · · · · · · · · · · · · 33	pyloric gland · · · · · · · · · · · · · · · · 105	renal · · · · · · · · · 14, 42, 45, 46, 122
protoplasm · · · · · · · · · · · · · · · · · · 50	pyloric insuffiency · · · · · · · · · · · 106	renal calculi · · · · · · · · · · · · · · · · 125
protozoa · · · · · · · · · · · · · · · · · · · 162	pyloric part · · · · · · · · · · · · · · · · · 106	renal calculus · · · · · · · · · · · · · · 123
protozoan · · · · · · · · · · · · · · · · · · 162	pyloric stenosis · · · · · · · · · · · · · · 106	renal corpuscle · · · · · · · · · · · · · 122
protozoans · · · · · · · · · · · · · · · · · 162	pylorospasm · · · · · · · · · · · · · · · · 106	renal failure · · · · · · · · · · · · · · · · 123
protozoon · · · · · · · · · · · · · · · · · · 162	pylorus · · · · · · · · · · 105, 106, 119	renal pelvis · · · · · · · · · · · · · · · · 122
protozoons · · · · · · · · · · · · · · · · · 162	pyocyst · 44	renomegaly · · · · · · · · · · · · · · · · 123
proximal · 11	pyogenesis · · · · · · · · · · · · · · · · · · 44	repeat breeder · · · · · · · · · · · · · 136
proximal interphalangeal joints of manus · · 61	pyometra · · · · · · · · · · · · · · · · 44, 129	replication · · · · · · · · · · · · · · · · · · 53
proximal phalanx · · · · · · · · · 58, 62	pyometritis · · · · · · · · · · · · · · · · · · 44	resorption · · · · · · · · · · · · · · · · · 161
proximal tubule · · · · · · · · · · · · · 122	pyopoiesis · · · · · · · · · · · · · · · · · · 44	respiration · · · · · · · · · · · · · · · · · · 90
pseudoarthrosis · · · · · · · · · · · · · · 63	pyorrhea · · · · · · · · · · · · · · · · 35, 44	respiratory · · · · · · · · · · · · · · · · · · 90
pseudomania · · · · · · · · · · · · · · · · 34	pyosalpinx · · · · · · · · · · · · · · · · · 128	respiratory acidosis · · · · · · · · · · 90
pseudopregnancy · · · · · · · · · · · 137	pyothorax · · · · · · · · · · · · · · · · 44, 92	retained corpus luteum · · · · · · 127
psychiatrist · · · · · · · · · · · · · · · · · 167	pyriform · 32	retained placenta · · · · · · · · · · · · 137
psychiatry · · · · · · · · · · · · · · · · · 167	pyromania · · · · · · · · · · · · · · · · · · · 34	retained testis · · · · · · · · · · · · · · 132
psychologist · · · · · · · · · · · · · · · 167	pyuria · · · · · · · · · · · · · · · · · · 44, 125	retention of corpus luteum · · · · 127
psychology · · · · · · · · · · · · · · · · · 167	**Q**	reticular fiber · · · · · · · · · · · · · · · · 52
ptyalism · 99		reticulitis · · · · · · · · · · · · · · · · · · · 108
ptyalography · · · · · · · · · · · · · · · · 99	quadriceps femoris · · · · · · · · · · · 26	reticuloendotheliosis virus · · · · 159
ptyalolith · 99	quadriplegia · · · · · · · · · · · · · 26, 36	reticulum · · · · · · · · · · · · · · 107, 108
pubis · 58	**R**	retina · · · · · · · · · · · · · · 43, 148, 149
public health · · · · · · · · · · · · · · · 167		retinal · · · · · · · · · · · · · · · · · · 43, 149
public health administrator · · · · 167	rachitis · 57	retinitis · · · · · · · · · · · · · · · · · · 43, 150
public health specialist · · · · · · · · 167	radiologist · · · · · · · · · · · · · · · · · 167	retinopathy · · · · · · · · · · · · · · 35, 150
pull out a tooth · · · · · · · · · · · · · 100	radiology · · · · · · · · · · · · · · · · · · 167	retinopexy · · · · · · · · · · · · · · · · · · 39
pulled muscle · · · · · · · · · · · · · · · 66	radiotherapy · · · · · · · · · · · · · · · · · 39	retractor bulbi muscle · · · · · · · · 148
pulmonary · · · · · · · · · · · · 14, 42, 95	radius · · · · · · · · · · · · · · · · 12, 56, 58	retro-phyryngeal lymph node · · 82
pulmonary artery · · · · · · · · · 71, 95	ram · 16	retrograde cystography · · · · · · 124
pulmonary edema · · · · · · · · · · · · 95	reactivate · · · · · · · · · · · · · · · · · · · 20	retrograde pyelography · · · · · · 123
pulmonary emphysema · · · · · · · 95	receptor · · · · · · · · · · · · · · · · · · · 141	retrograde urography · · · · · · · · 125
pulmonary incompetence · · · · · · 72	recombinant · · · · · · · · · · · · · · · · · 20	rhabdomyolysis · · · · · · · · · · · · · · 66
pulmonary infarct · · · · · · · · · · · · 95	recovery of activity · · · · · · · · · 168	rheumatism · · · · · · · · · · · · · · · · · 33

rheumatoid arthritis ·········· 86	sacroiliac joint ············· 62	seminiferous tubule ·········· 135
rhinal ··············· 15, 42, 45	sacrum ·················· 57	semino-vesiculitis ············ 134
rhinalgia ············· 45, 152	sagittal ·················· 8	seminoma ············· 132, 135
rhinitis ··········· 42, 91, 152	saliva ················ 98, 99	semitendinous muscle ········ 20
rhinodynia ············ 45, 152	salivary ················· 99	sensory nerve ············· 140
rhinoplasty ················ 37	salivary glands ············ 98	sensory neuron ············· 140
rib ·················· 12, 15	salivate ················· 99	septicemia ················ 78
ribonucleic acid (RNA) ······· 53	salivation ················ 99	serologist ················ 167
ribosomal RNA (rRNA) ······ 53	salpingectomy ············· 128	serology ················· 167
ribosome ················ 51	salpingian ············ 42, 128	serotherapy ··············· 39
ribs ··················· 56	salpingitis ············ 42, 128	Sertoli cell tumor ········ 132, 145
rickets ·················· 57	salpingo-oophoritis ·········· 128	serum ··················· 80
right and left bronchi ········· 93	salpingoma ··············· 128	sesamoid ················· 32
right and left hepatic ducts ···· 117	salpingostomy ·············· 38	sexual reproduction ·········· 164
right atrio-ventricular valve ···· 71	sanguiferous ··············· 78	shaft ··················· 59
right atrium ··············· 71	sanguineous ··············· 78	she-cat ·················· 16
right bundle branches ········ 73	SA node ················· 73	sheep ··················· 16
right limbs················· 73	sarcolemma ··············· 64	short period estrus ·········· 136
right lung ············· 93, 94	sarcomere ················ 64	shoulder ················· 12
right ventricle ··········· 14, 71	sarcoplasm ················ 64	shoulder joint ··········· 61, 62
right ventricular ············ 14	sarcoplasmic reticulum ········ 64	shoulder joint region ·········· 10
rigor mortis ··············· 161	sarcoptidae ··············· 164	sialism ·················· 99
RNA virus ················ 159	scapula ············ 12, 56, 58	sialismus ················· 99
roaring ·················· 91	*Schistosoma japonicum* ······ 162	sialodochitis ··············· 99
rodenticide ················ 34	schistosomiasis············· 162	sialogram ················· 99
root of tooth ·············· 100	schistosomiasis japonicum ···· 162	sialolith ·················· 99
rostral ················· 8, 9	Schwann cell ·············· 141	sialorrhea ················· 99
rotator muscle ·············· 66	sciatic nerve··············· 46	silent estrus ··············· 136
rough endoplasmic reticulum ·· 51	scintigram ················ 31	silent heat ················ 136
round worm ··············· 163	sclera ··············· 148, 149	sinoatrial node ············· 73
rumen ·············· 107, 108	scleritis ················· 150	sinusitis·················· 91
rumen acidosis ············ 108	sclerosis ················· 35	skeletal muscle ············· 65
rumen alkalosis············ 108	screral·················· 149	skeleton of forearm ··········· 58
rumen putrefaction ·········· 108	scrota ·················· 134	skeleton of leg·············· 58
rumenotomy ·············· 108	scrotal ·················· 134	skeleton of manus ··········· 58
rumens ················· 107	scrotum ··········· 132, 133, 134	skeleton of pes ············· 58
rumina ················· 107	scurvy ··················· 78	skeleton of thigh ············ 58
ruminal ················· 107	sebaceous epithelioma ······· 159	skeleton of upper arm ········ 58
ruminal impaction ··········· 108	sebaceous gland ············ 154	skin ··············· 43, 46, 154
ruminal parakeratosis ········ 108	secondary hair ············· 154	skull ··················· 56
ruminal tympany ············ 108	secretory vesicle ············ 51	small bowel ··············· 110
ruminant ················· 107	sedative ················· 168	small intestine ·· 98, 110, 112, 113
ruminate ················· 107	semen··················· 135	small pastern bone ········ 58, 62
rumination ················ 107	semicircular ducts············ 151	smooth endoplasmic reticulum ·· 51
ruminitis ················· 108	semilunar fold of conjunctiva ··· 149	smooth muscle ············· 65
rupture of bladder ··········· 124	semilunar valves ············ 71	snore···················· 91
	semimembranous muscle ······ 20	snout ··················· 10
S	seminal ················· 135	sole ················ 155, 156
	seminal vesicle ············· 133	solitary lymphatic nodule ······ 82
sacral vertebrae ············ 57	seminal vesiculitis············ 134	somatic nervous system ······ 140

187

somatostatin 119	stomach acid 106	swine 16
sow 16	stomach cancer 105	swine atrophic rhinitis 91, 152
spasm 143	stomach flu 105	swine transmissible gastroenteritis .. 110
spasm of esophagus 103	stomach pump 105	sympathetic ganglia 143
spay 137	stomach tube 105	sympathetic nerve 140
spectrometer 31	stomach ulcer 105	sympathetic trunk 143
spectrometry 31	stomach worm 105	symphysis 20
sperm 135	stomachache 105	symptom 20
spermatic 135	stomachic 105	synapse 20
spermatic cord 132, 134	stomatalgia 45, 99	synchronous diaphragmatic flutter in horses 93
spermatogenesis 32, 135	stomatic 99	syndrome 20
sphincter muscle 66	stomatitis 99	syngamy 164
spinal 143	stomatodynia 45, 99	syngraft 20
spinal arachnoid 142	stool examination 168	synovia 63
spinal cord 140, 141, 142, 143	strain 66	synovial bursa 63
spinal dura matter 142	striated muscle 65	synovial fluid 63
spinal ganglion 143	stroke 143	synovial membrane 63
spinal pia mater 142	stroma 158	synovial sarcoma 159
spinous epithelial cell 155	strongyloides 163	synovial sarcoma 63
spinous layer 155	strongyloidiasis 163	synovial sheath of tendon 65
spinous process 57	struma 145	synovioma 63
splanchnic 104	sty 150	systemic lupus erythematosus .. 86
splanchnology 104	subarachnoid cistern 142	systole 73
spleen 13, 14, 84, 119	subarachnoid hemorrhage 24	systolic murmur 73
splenectomy 84	Subclass acari 164	
splenectopia 84	subcutaneous injection 24	T
splenectopy 84	subcutaneous tissue 154	
splenic 14, 84	subcutaneous (S.C.) 168	T cell 85
splenitis 84	subluxation 59, 170	tachycardia 73
splenomegaly 84	substantia nigra 25	*Taenia solium* 163
spondylosis 143	superacute 24	Taenia solium taeniasis 163
spondylosis deformans 63	superficial 11, 24	taeniasis 163
sprain 63	superficial cervical lymph node .. 82	tail 12, 15
sputum examination 168	superior 9	tail of muscle 65
squamous cell carcinoma 159	supination 66	tapetum 148
stallion 16	supinator muscle 66	tapeworm 162, 163
stapes 151	supporting cell 153	tapeworm disease 163
statistics 33	suppository 114	tarsal 15
steer 16	suppurate 44	tarsal bones 12, 56, 58
steosarcoma 159	suppuration 44	tarsal joint 61
sternum 12, 56	suppurative inflammation 161	tastant 153
stertorous respiration 91	suppurative myositis 66	taste bud 153
stethoscope 30, 73	supraorbital nerve 24	taste cell 153
stethoscopy 30	suprarenal gland 144	teat 131
stifle 15	supraspinatus muscle 24	teat canal 131
stifle joint 62	supraventricular tachycardia 74	teeth 14, 45, 98, 100, 102
stillbirth 137	surgeon 167	temporary teeth 102
stirup 151	surgery 167	temporomandibular joint .. 61, 62
stomach .. 13, 14, 42, 45, 98, 105, 109, 113, 119	suspensory ligment of udder .. 131	tendinous cord 71
	suture 61	tendinous sheath 65

tendon ······ 65	tibia ······ 12, 56, 58	trichomonas ······ 162
tension ······ 66	tic ······ 66	trichomoniasis ······ 162
teratologist ······ 167	tick ······ 164	trichuriasis ······ 163
teratology ······ 137, 167	tight junction ······ 50	trichuris ······ 163
terminal bone ······ 58	toe ······ 15	tricuspid incompetence ······ 72
testalgia ······ 46, 132	tomograph ······ 31	tricuspid insufficiency ······ 72
testes ······ 132	tongue ······ 14, 42, 45, 90, 153	tricuspid regurgitation ······ 72
testicular ······ 14, 46, 132	tonsil ······ 42, 83	tricuspid valve ······ 71
testicular hypoplasia ······ 132	tonsillar ······ 42, 83	trigeminal nerve ······ 140
testicular tumor ······ 132	tonsillary ······ 42, 83	trochlear nerve ······ 140
testis ······ 13, 14, 46, 132, 133, 144	tonsillectomy and adenoidectomy ······ 83	truncal ······ 15
testitis ······ 132	tonsillitis ······ 42, 83, 99	trunk ······ 15
tetany ······ 143	tooth ······ 14, 100, 102	trypanosoma ······ 162
tetracycline ······ 26	toothache ······ 100	trypanosomiasis ······ 162
tetraplegia ······ 36	torsion of abomasum ······ 108	tubectomy ······ 128
tetrapod ······ 26	toxicologist ······ 167	tubo-ovaritis ······ 128
thalamus ······ 142	toxicology ······ 167	tubulo-interstitial nephritis ······ 123
theileria ······ 162	toxoplasma ······ 162	tumors of the liver ······ 115
theileriasis ······ 162	toxoplasmosis ······ 162	tunica externa ······ 76
therapeutic ultrasound ······ 169	trachea ······ 42, 45, 90, 93, 94	tunica media ······ 76
thermograph ······ 31	tracheae ······ 94	tup ······ 16
thermometer ······ 31	tracheal ······ 42, 45, 94	tusk ······ 102
thermometry ······ 31	tracheal bronchus ······ 93	tusks ······ 102
thermotherapy ······ 39, 169	tracheal collapse ······ 94	twisting of the intestines ······ 110
thigh ······ 12, 15	trachealgia ······ 45	twitch ······ 66
third eyelid ······ 149	tracheitis ······ 42, 94	tympanic cavity ······ 151
third ventricle ······ 142	tracheo-bronchial lymph node ······ 82	tympanic membrane ······ 151
thoraces ······ 92	tracheostomy ······ 38, 94	typhlitis ······ 113
thoracic ······ 15, 92	trachestomy ······ 94	
thoracic cavity ······ 92	trachitis ······ 94	# U
thoracic duct ······ 79, 82, 92	transcription ······ 53	udder ······ 130
thoracic limb ······ 15	transfer RNA (tRNA) ······ 53	ulna ······ 12, 56, 58
thoracic vertebrae ······ 57	transfusion ······ 22, 168	ultrasonogram ······ 31
thoracostomy tube ······ 92	transgenic ······ 53	ultrasonography ······ 70, 168
thorax ······ 15, 92	translation ······ 53	umbilical ······ 15
thoraxes ······ 92	transmembrane ······ 22	umbilical cord ······ 136, 137
throat ······ 12	transplantation ······ 86	umbilicus ······ 15
thrombocyte ······ 80	transplantation of the heart ······ 72	unconscious ······ 19
thrombolysis ······ 35	transverse colon ······ 109	undescended testis ······ 132
thrombopoiesis ······ 32	transverse process ······ 57	unguiculate bone ······ 58
thrombosis ······ 76	trasal region ······ 10	ungulata ······ 58
thumps ······ 93	traumatic reticulitis ······ 108	uniculus spermaticus ······ 132
thymi ······ 83	traumatic reticuloperitonitis ······ 108	unilateral ······ 26
thymic ······ 83	traumatic splenitis ······ 84	unmyelinated fiber ······ 19
thymoma ······ 83	trematode ······ 162	upper arm ······ 15
thymus ······ 83	triceps brachii ······ 26	uremia ······ 125
thymuses ······ 83	trichinella ······ 163	ureter ······ 122, 123, 127, 132, 133
thyroid cartilage ······ 90	trichinellosis ······ 163	ureteral ······ 123
thyroid gland ······ 144	trichinosis ······ 163	ureteral calculi ······ 125
thyroiditis ······ 145	trichoepithelioma ······ 159	ureteralgia ······ 123

ureterectomy ······················ 37	vaginitis ··············· 43, 130	viscus ····················· 104
ureteritis ······················· 123	vaginodynia············· 46, 130	vitreous body ················ 148
ureterocystostomy ············ 123	vaginohysterectomy ·········· 130	vocal cord ···················· 91
ureterolith ····················· 125	vaginoperineorrhapy ··········· 38	vocal muscle ·················· 91
ureterotomy ···················· 37	vaginoplasty····················· 37	voluntary muscle ··············· 65
urethra ··13, 42, 46, 122, 124, 127, 132, 133	vaginoscope ·················· 30	volvulus ····················· 110
	vaginoscopy···················· 30	vomeronasal organ ··········· 152
urethral ················ 42, 46, 124	vagus nerve ················· 140	vomiting ····················· 99
urethral calculi ················ 125	valvule ······················· 76	vulva ························ 127
urethral stones ················ 125	vas deferens ················· 133	
urethralgia ················ 46, 124	vascular ······················ 75	**W**
urethritis ·················· 42, 124	vasculitis ····················· 75	waist ························· 15
urethrodynia ··············· 46, 124	vasectomy ················ 37, 134	wall······················ 155, 156
urinary ······················· 125	vasitis ······················ 134	weight loss ·················· 168
urinary bladder ·· 13, 14, 122, 124, 127, 132, 133	vein ·························· 76	whiplash injury················ 63
	vena cava caudalis ········ 79, 94	white blood cell ··············· 80
urinary retention ············· 125	venostasis ···················· 76	white corpuscle ··············· 80
urinary stone ················· 125	venous ······················· 76	white line ··············· 155, 156
urinate ······················ 125	ventral························ 8, 9	white matter ················· 141
urinating ····················· 46	ventral rectus muscle ········ 148	white pulp···················· 84
urine ························ 125	ventricle ······················ 71	white substance ·············· 141
urine examination ············ 168	ventricles ················ 71, 142	woman ······················· 16
urocystitis····················· 42	ventricular ···················· 71	womb ······················· 129
urodynia ····················· 46	ventricular fibrillation ·········· 73	women ······················· 16
urologist ····················· 167	ventricular premature complex ·· 74	wrist ························· 15
urology ······················ 167	ventricular septal defect········ 72	wrist joint ···················· 61
urovagina ···················· 130	ventricular tachycardia ········ 74	
uteri························· 129	ventriculocordectomy ·········· 91	**X**
uterine ··············· 43, 46, 129	venule ······················· 76	X-ray ························ 168
uterine body ················· 129	vertebra ······················ 12	xanthemia ···················· 25
uterine hernia ················ 129	vertebrae ····················· 56	xanthine ······················ 25
uterine horn ·················· 129	vertebral column ··············· 57	xanthoma ················ 25, 160
uterine inertia ················ 137	vertebrate foramen ············ 57	xanthopsia ···················· 25
uterine prolapse ·············· 129	vesica biliaris ················· 117	
uterine torsion ··········· 129, 137	vesical ······················ 124	**Z**
uteritis ··················· 43, 129	vesical calculi ················ 125	zonula occludens ············· 50
uterus ············ 43, 46, 127, 129	vesical calculus··············· 124	zoologist····················· 167
	vessel ························ 75	zoology ······················ 167
V	vessels ······················ 154	zoonosis ······················ 35
vaccination ··················· 168	vestibule of vagina ············ 127	
vaccine ·················· 86, 168	vestibulocochlear nerve ······ 140	
vacuolar degeneration ········ 160	vestibulocochlear organ ······ 151	
vagal indigestion ············· 108	viral hepatitis ················· 115	
vagina ······ 13, 43, 46, 127, 130	viremia ······················ 78	
vaginal ················ 43, 46, 130	virologist····················· 167	
vaginal cyst ·················· 130	virology ····················· 167	
vaginal hysterectomy ········· 130	visceral ····················· 104	
vaginal plug ·················· 130	visceral layer of the serous pericardium 74	
vaginal prolapse ·············· 130	visceromotor neuron ········· 104	
vaginal stenosis ·············· 130	viscerosensory ··············· 104	

謝辞

　本書執筆にあたり、多くの先生方にお世話になりました。Dr. Steve Adams（Purdue University、第1章：Hoovesなどの指導）、及川正明先生（北里大学、第1章：馬の体部の名称）、Dr. Amie Koenig（University of Georgia、第2、3章：Prefixes、Suffixesの校閲）、左近允巌先生（北里大学、第5章：筋骨格系の助言）、Dr. Henry Green（Purdue University、第6章：Cardiologyの指導と校閲）、Dr. Brenda Austin（Purdue University、第6章：Lymphatic Organsの助言）、Dr. Gary C. Lantz（Purdue University、第8章：Dentistryの指導）、三浦弘先生（北里大学、第9章：臨床繁殖学の校閲）、Dr. Patrick Hensel (University of Georgia、第11章：Dermatologyの指導、校閲)、Dr. Sulma Mohammed（Purdue University、第12章：Oncologyの校閲）、朴天鎬先生（北里大学、第12章：病理学の校閲）、工藤上先生（北里大学、第12章：寄生虫学の助言と資料）、藤崎幸蔵先生（鹿児島大学、第12章：寄生虫学の指導と校閲）、Dr. Tomohiro Inoue（Purdue University、Anesthesiologyの指導）、Dr. Stephanie Inoue（Pets and Vets as Partners（Animal Hospital）、Rehabilitationの指導）、本田剛獣医師（本田動物病院、臨床用語の助言）、Mr. Steven DeBonis（St. Ursula Gakuin、英語の校閲）、Mr. William Dantona（California State University、英語の校閲）、谷口和之（岩手大学、全体の校閲）、ならびに、情熱と専門知識を駆使して本書を編集して下さった緑書房／チクサン出版社の沼田利恵氏と後藤瑞枝氏に心から感謝申し上げます。

[参考図書]

1. 新編　家畜比較解剖図説、上巻、下巻、加藤嘉太郎、山内昭二著、養賢堂、2003
2. 犬の解剖アトラス　日本語版（第2版）　林良博、橋本善春監修、学窓社、2004
3. Atlas of Bovine Anatomy. Chris Pasquini, Sudz Publishing, TX, USA, 1983
4. Anatomy of Domestic Animals Systemic and Regional Approach. 11th edition. Pasquini, Tom Spurgeon, Chris Pasquini, Sudz Publishing, TX, USA, 2003
5. Atlas of Topographic Anatomy of the Domestic Animal. Peter Popesko, W.B.Saunders, 1971
6. 動物病理学総論（第2版）、板倉智敏、後藤直彰編、文永堂出版、2001
7. スモールアニマル・サージェリー（第3版日本語版）、Theresa Welch Fossum著、インターズー、2008
8. 新版 獣医臨床寄生虫学、獣医臨床寄生虫学編集委員会編、文永堂出版、1995
9. ステッドマン医学大辞典　改訂第6版 ステッドマン医学辞典編集委員会編、メジカルビュー社、1998
10. Oxford Advanced Learner's Dictionary, 7th edition, Oxford University press, 2005
11. Webster's New World Dictionary College edition. Prentice Hall General, 1988
12. 医学語源散策　岩槻賢一著、医学図書出版、2000
13. これだけは知っておきたい医学ラテン語　二宮陸雄著　講談社、1986
14. 医学ラテン語のてびき　川名悦郎、原田幸雄、和田廣著、同学社、1984

[著者紹介]

谷口 和美 (たにぐち かずみ)
Kazumi Taniguchi, D.V.M, Ph.D

獣医師、農学博士。東京都生まれ。
東京大学農学部畜産獣医学科卒業、東京大学大学院修了。
日本Roche研究所、岩手医科大学医学部解剖学講座、北里大学獣医学部解剖学研究室を経て、現在、岩手大学非常勤講師。
2001年－2002年、米国Pennsylvania州、Philadelphia市、Monell Chemical Senses Centerに留学。

パーフェクト獣医学英語

2009年5月1日　第1刷発行
2020年4月20日　第4刷発行

- ■著　者／谷口和美
- ■発行者／森田 猛
- ■発　行／チクサン出版社
- ■発　売／株式会社 緑書房
  〒103-0004
  東京都中央区東日本橋3丁目4番14号
  TEL03-6833-0560
  http://www.pet-honpo.com

- ■印刷所／図書印刷

落丁・乱丁本は、弊社送料負担にてお取り替えいたします。
©Kazumi Taniguchi
ISBN978-4-88500-663-0　　Printed in Japan

本書の複写にかかる複製、上映、譲渡、公衆送信（送信可能化を含む）の各権利は株式会社緑書房が管理の委託を受けています。

JCOPY 〈(一社)出版者著作権管理機構 委託出版物〉
本書を無断で複写複製（電子化を含む）することは、著作権法上での例外を除き、禁じられています。本書を複写される場合は、そのつど事前に、(一社)出版者著作権管理機構（電話 03-5244-5088、FAX 03-5244-5089、e-mail：info@jcopy.or.jp）の許諾を得てください。
また本書を代行業者等の第三者に依頼してスキャンやデジタル化することは、たとえ個人や家庭内での利用であっても一切認められておりません。

- ■イラスト／谷口和美、パント末吉、Dr. Carol Bain、末武創一
- ■カバー・本文デザイン／Design PANTO'S